Adding and subtracting integers

Adding a **negative** number **or** subtracting a **positive** number will have the **same result**.

$$3 + -5 = -2$$
$$3 - +5 = -2$$

Go down by 5.

Use a number line to visualise the answer.

Adding a negative number means subtract. →

Subtracting a negative number means add. →

Adding a positive number **or** subtracting a negative number will have the **same result**.

$$-1 + +4 = +3$$
$$-1 - -4 = +3$$

Go up by 4.

+ +	means	+
+ −	means	−
− +	means	−
− −	means	+

Multiplying and dividing integers

Look at these examples.

Multiplying a negative number by a positive number always gives a negative answer.

$$-5 \times +3 = -15$$
$$+5 \times -3 = -15$$

Multiplying two positive numbers **or** multiplying two negative numbers always gives a positive answer.

$$+4 \times +3 = +12$$
$$-4 \times -3 = +12$$

The same rules work for division.

$$+10 \div -5 = -2$$
$$-10 \div -5 = +2$$

This table summarises the rules:

+	× or ÷	+	=	+
+	× or ÷	−	=	−
−	× or ÷	+	=	−
−	× or ÷	−	=	+

A positive number multiplied by a negative number gives a negative answer.

A negative number multiplied by a negative number gives a positive answer.

KEYWORDS

Integer ➤ An integer is a whole number; it can be positive, negative or zero.

Positive ➤ A number above zero.

Negative ➤ A number below zero.

Use of symbols

Look at the following symbols and their meanings.

Symbol	Meaning	Examples
>	Greater than	5 > 3 (5 is greater than 3)
<	Less than	−4 < −1 (−4 is less than −1)
⩾	Greater than or equal to	$x \geqslant 2$ (x can be 2 or higher)
⩽	Less than or equal to	$x \leqslant -3$ (x can be −3 or lower)
=	Equal to	2 + +3 = 2 − −3
≠	Not equal to	$4^2 \neq 4 \times 2$ (16 is not equal to 8)

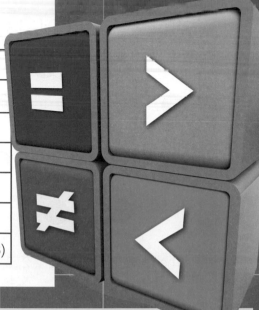

Place value

Look at this example.

Given that 23 × 47 = 1081, work out 2.3 × 4.7

The answer to 2.3 × 4.7 must have the digits 1 0 8 1 ← Do a quick estimate to find where the decimal point goes.

2.3 is about 2 and 4.7 is about 5. Since 2 × 5 = 10, the answer must be about 10.

Therefore 2.3 × 4.7 = 10.81

Write the following symbols and numbers on separate pieces of paper.

     0

Arrange them to form a correct calculation.

How many different calculations can you make? For example:

1. Calculate the following:
   (a) −5 − −8
   (b) −2 + −6
   (c) −7 + −3 − −5
2. Calculate the following:
   (a) −12 × −4
   (b) 24 ÷ −3
   (c) −3 × −4 × −5
3. State whether these statements are true or false.
   (a) 6 < 3
   (b) −4 > −5
   (c) 2 + −3 = 2 − +3
4. Given that 43 × 57 = 2451, calculate the following:
   (a) 4.3 × 0.57
   (b) 430 × 570
   (c) 2451 ÷ 5.7

Highest common factor

The highest common factor (HCF) is the highest **factor** shared by two or more numbers.

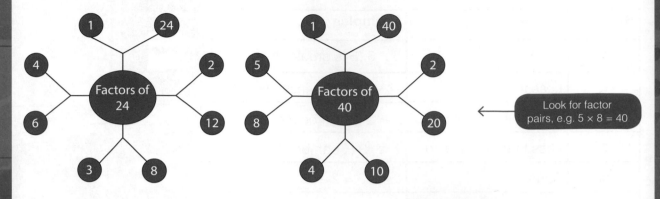

Look for factor pairs, e.g. 5 × 8 = 40

The factors of 24 are 1, 2, 3, 4, 6, ⑧, 12 and 24.

The factors of 40 are 1, 2, 4, 5, ⑧, 10, 20 and 40.

Remember – every number has 1 and itself as a factor.

8 is the HCF of 24 and 40.

Lowest common multiple

The lowest common multiple (LCM) is the lowest **multiple** shared by two or more numbers.

The first seven multiples of 5 are: 5, 10, 15, 20, 25, ㉚, 35, …

The first seven multiples of 6 are: 6, 12, 18, 24, ㉚, 36, 42, …

The first multiple of any number is itself.

30 is the LCM of 5 and 6.

Prime factors

Numbers can be expressed as the **product** of their **prime factors**. The product of the prime factors can be shown in **index form** where appropriate.

Write 24 as a product of its prime factors.

Use a factor tree to help. Start with the smallest **prime** number that divides exactly into 24 and continue in order until 1 is reached.

2

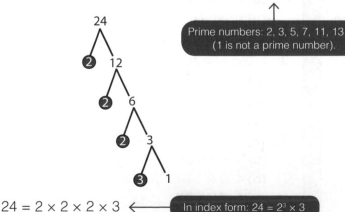

Prime numbers: 2, 3, 5, 7, 11, 13, … (1 is not a prime number).

$24 = 2 \times 2 \times 2 \times 3$ ← In index form: $24 = 2^3 \times 3$

HCF and LCM using prime factors

You can find the HCF of two numbers by multiplying their shared prime factors.

You can find the LCM of the two numbers by multiplying the HCF with the rest of the prime factors.

Find the HCF and LCM of 24 and 42.

$24 = 2 \times 2 \times (2 \times 3)$

$42 = (2 \times 3) \times 7$

This can be presented in a Venn diagram.

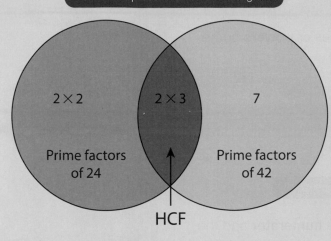

| 2 × 2 | 2 × 3 | 7 |

Prime factors of 24   Prime factors of 42

HCF

The **HCF** of 24 and 42 = 2 × 3 = **6**

The LCM of 24 and 42 = 6 × the rest of the prime factors

The **LCM** = 6 × 2 × 2 × 7 = **168**

Write the following prime numbers on separate pieces of paper.

| 2 | 2 | 3 | 3 | 5 | 7 |

Draw a Venn diagram on a piece of paper and place the prime numbers anywhere in the diagram (at least one in each of the three sections).

Find the product of each of the sections. You will now have two numbers and the HCF of those numbers. Find the LCM of your two numbers.

Repeat by rearranging your prime factors.

KEYWORDS

Factor ➤ A number which divides exactly into another number.

Multiple ➤ A number within the same times tables as another number.

Product ➤ The number or quantity obtained by multiplying two or more numbers together.

Prime factor ➤ A factor which is also prime.

Index form ➤ A number written using powers.

Prime ➤ A number with **exactly** two factors, 1 and itself.

---

1. (a) Find all the factors of 12 and 15.
   (b) Find the HCF of 12 and 15.
2. (a) List the first 8 multiples of 7 and 5.
   (b) Find the LCM of 7 and 5.
3. Answer these.
   (a) Write the following numbers as a product of their prime factors: 28 and 98.
   (b) Write your answers to part (a) in index form.
   (c) Find the HCF and the LCM of 28 and 98.

Using Bidmas

Bidmas helps you to remember the correct order of a calculation.

Brackets

Indices

Division

Multiplication

Addition

Subtraction

> Start with brackets and then follow this order. Remember that multiplying and dividing comes before adding and subtracting.

Calculate $4 + 3 \times 6$ ← Multiply first and then add.

$4 + 18 = 22$

Remember that brackets followed by indices or **powers** come before any of the four operations.

Calculate $56 - 3^2(4 + 2)$

$56 - 3^2(6)$ ← Brackets

$56 - 9 \times 6$ ← Powers

$56 - 54 = 2$ ← Multiplication then subtraction

In the next example you must calculate the **numerator** and the **denominator** separately before dividing.

$$\frac{8 + 6}{(8 - 3) \times 20} = \frac{14}{5 \times 20} = \frac{14}{100} = 0.14$$

Reciprocals

The **reciprocal** of n is $1 \div n$ or $\frac{1}{n}$

Therefore the reciprocal of 5 is $1 \div 5$ or $\frac{1}{5}$

The reciprocal of $\frac{2}{3}$ is $1 \div \frac{2}{3} = 1 \times \frac{3}{2} = \frac{3}{2}$

> To find the reciprocal you can simply turn the fraction upside down.

Adding and subtracting fractions

You can only add or subtract fractions which have a common denominator.

$\frac{1}{5} + \frac{3}{5} = \frac{1 + 3}{5} = \frac{4}{5}$ ← Keep the denominator the same.

If the denominators are different then find equivalent fractions with the same denominator.

$\frac{2}{3} - \frac{1}{4}$ ← The LCM of 3 and 4 is 12.

× by 4   × by 3

$\frac{8}{12} - \frac{3}{12}$ $= \frac{8 - 3}{12} = \frac{5}{12}$ ← Subtract as normal.

Module 3   Operations

Multiplying and dividing fractions

When multiplying fractions, simply multiply the numerators and then multiply the denominators.

$$\frac{2}{3} \times \frac{3}{5} = \frac{2 \times 3}{3 \times 5} = \frac{6}{15} = \frac{2}{5}$$ ← Don't forget to simplify.

When dividing fractions, you don't actually divide at all. Find the reciprocal of the second fraction and then multiply instead.

multiply instead of divide

$$\frac{2}{3} \div \frac{4}{5} = \frac{2}{3} \times \frac{5}{4} = \frac{2 \times 5}{3 \times 4} = \frac{10}{12} = \frac{5}{6}$$

reciprocal of 2nd fraction

$$\frac{2}{3} \div \frac{4}{5} = \frac{5}{6}$$

Calculating with mixed numbers

When multiplying or dividing with mixed numbers, it's best to convert them to improper fractions.

$$2\frac{1}{4} \times 1\frac{2}{3} = \frac{9}{4} \times \frac{5}{3} = \frac{9 \times 5}{4 \times 3} = \frac{45}{12} = \frac{15}{4} = 3\frac{3}{4}$$

change to improper fractions

$$2\frac{1}{4} \times 1\frac{2}{3} = 3\frac{3}{4}$$

When adding or subtracting mixed numbers, you should deal with the fraction part separately.

$$2\frac{1}{4} \quad + \quad 3\frac{1}{3}$$

$$2 + 3 = 5 \qquad \frac{1}{4} + \frac{1}{3} = \frac{3}{12} + \frac{4}{12} = \frac{7}{12}$$

$$2\frac{1}{4} + 3\frac{1}{3} = 5\frac{7}{12}$$

Fold a piece of A4 paper in half three times. Open it out and shade $\frac{3}{8}$ of it.

Fold the paper back in half five times. Open it out and shade in $\frac{9}{32}$ (make sure you shade nine unshaded areas).

Count the total number of shaded squares to find the answer to $\frac{3}{8} + \frac{9}{32}$.

1. Calculate the following:
   (a) $6 + 3 \times 4$
   (b) $(7 - 4) \times (6 + 2)$
   (c) $(3 + 4^2) \times 2$
   (d) $\dfrac{5^2 + 3}{2 \times \sqrt{49}}$

2. Find the reciprocal of:
   (a) $\frac{2}{5}$
   (b) $\frac{1}{7}$
   (c) 5
   (d) 0.25

3. Work out:
   (a) $\frac{4}{5} - \frac{3}{4}$
   (b) $2\frac{1}{4} + 3\frac{3}{8}$
   (c) $\frac{4}{7} \div \frac{3}{5}$
   (d) $2\frac{2}{3} \times 1\frac{3}{4}$

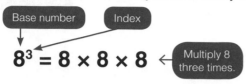

Index form and roots

The index number tells you how many times the base number is multiplied.

Base number   Index

$8^3 = 8 \times 8 \times 8$ ← Multiply 8 three times.

Base number   Index

$4^5 = 4 \times 4 \times 4 \times 4 \times 4$ ← Multiply 4 five times.

The following table shows you how to calculate **roots**.

Roots	How to say it	Reasoning
$\sqrt{25} = 5$	Square root of 25 equals 5	Because $5 \times 5 = 25$
$\sqrt[3]{8} = 2$	Cube root of 8 equals 2	Because $2 \times 2 \times 2 = 8$
$\sqrt[4]{81} = 3$	Fourth root of 81 equals 3	Because $3 \times 3 \times 3 \times 3 = 81$
$\sqrt[5]{32} = 2$	Fifth root of 32 equals 2	Because $2 \times 2 \times 2 \times 2 \times 2 = 32$

4

Index laws ← Index laws only apply when the base numbers are the same.

Multiplying ⟶ Add powers to give final index.

$4^2 \times 4^3 = 4^{2+3} = 4^5$ $a^2 \times a^{-3} \times a^6 = a^{2-3+6} = a^5$ ← Normal rules apply for negative numbers.

Dividing ⟶ Subtract powers to give final index.

$5^7 \div 5^3 = 5^{7-3} = 5^4$ $b^9 \div b = b^{9-1} = b^8$ ← If no power is given, assume it is 1.

Raising one power to another power ⟶ Multiply powers to give final index.

$(3^2)^4 = 3^{2 \times 4} = 3^8$ $(c^2)^{-3} = c^{2 \times -3} = c^{-6}$ Also see page 21.

Zero and negative indices

These rules can be used to **evaluate** more difficult indices.

Rule	Example
$x^0 = 1$	$5^0 = 1$
$x^{-n} = \dfrac{1}{x^n}$	$6^{-2} = \dfrac{1}{6^2} = \dfrac{1}{36}$

Any non-zero number raised to the power of 0 equals 1.

Write down the reciprocal and change the sign of the index.

Fold a piece of paper in half six times and measure its thickness.

If it were possible, how many times would you need to fold it until it was thicker than 30cm? Or taller than your house? Or taller than the Eiffel Tower?

Open the piece of paper which was folded six times and count the rectangles. Consider why the total number of rectangles is $2^6 = 64$.

Standard form

Standard form is a useful way of writing very large or very small numbers using powers of 10.
A number written in standard form looks like this:

$$A \times 10^n$$

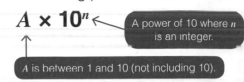

A power of 10 where n is an integer.

A is between 1 and 10 (not including 10).

Look at this example to write very large numbers in standard form:

$5\,000\,000 = 5 \times 1\,000\,000$ ← 1 million = 10^6

$ = 5 \times 10^6$

This is 5 000 000 in standard form.

$43\,000\,000$ is written as 4.3×10^7

Use all the non-zero digits.

Count all the digits after the 4 (1 + 6 = 7).

Know the following powers of 10:

$1000 = 10^3$

$100 = 10^2$

$10 = 10^1$

$1 = 10^0$

$0.1 = 10^{-1}$

$0.01 = 10^{-2}$

$0.001 = 10^{-3}$

Look at this example to write numbers less than 1 in standard form:

$0.00007 = 7 \div 100\,000$

$ = 7 \div 10^5$

$ = 7 \times 10^{-5}$

Index number is always negative for these types.

0.000638 is written as 6.38×10^{-4} ← The digits have moved four decimal places.

Calculating with standard form

When multiplying or dividing, deal with the powers separately.

$(4 \times 10^6) \times (8 \times 10^{-3}) = (4 \times 8) \times (10^6 \times 10^{-3}) = 32 \times 10^3$

Remember to put your answer back into standard form → $= 3.2 \times 10^4$

$(5 \times 10^7) \div (2 \times 10^6) = \dfrac{5 \times 10^7}{2 \times 10^6}$

$ = \dfrac{5}{2} \times \dfrac{10^7}{10^6}$

$ = 2.5 \times 10^1$ ← Subtract indices when dividing.

When adding or subtracting, convert the values to ordinary numbers first.

$(5 \times 10^5) - (4 \times 10^4) = 500\,000 - 40\,000 = 460\,000$

$ = 4.6 \times 10^5$

$(3.2 \times 10^{-3}) + (4.3 \times 10^{-4}) = 0.0032 + 0.00043$

$\phantom{(3.2 \times 10^{-3}) + (4.3 \times 10^{-4})} = 0.00363$

$\phantom{(3.2 \times 10^{-3}) + (4.3 \times 10^{-4})} = 3.63 \times 10^{-3}$

1. Write the following in index form.
   (a) $4 \times 4 \times 4$
   (b) $3 \times 3 \times 3 \times 3 \times 3$
   (c) $\dfrac{1}{2 \times 2 \times 2}$

2. (a) Simplify the following, leaving your answer in index form.
   (i) $3^4 \times 3^5$ (ii) $4^8 \div 4$
   (iii) $\dfrac{6^2 \times 6^8}{6^3}$
   (b) Evaluate the following.
   (i) 7^0 (ii) 5^{-3}
   (iii) $\sqrt[4]{16}$

3. (a) Write the following in standard form.
   (i) 6000 (ii) 2300
   (iii) 0.00678 (iv) 0.15
   (b) Calculate the following, leaving your answer in standard form.
   (i) $(3 \times 10^4) \times (6 \times 10^5)$
   (ii) $(8 \times 10^3) \div (4 \times 10^{-3})$
   (iii) $(5.1 \times 10^6) - (3.9 \times 10^5)$

Converting between fractions, decimals and percentages

Make sure you know these key fraction, percentage and decimal equivalences.

Fraction	Decimal	Percentage
$\frac{1}{3}$	0.333…	$33\frac{1}{3}\%$
$\frac{1}{4}$	0.25	25%
$\frac{2}{5}$	0.4	40%
$\frac{3}{10}$	0.3	30%
$\frac{5}{100}$	0.05	5%
$\frac{75}{100}$	0.75	75%
$\frac{3}{8}$	0.375	$37\frac{1}{2}\%$

$1 \div 4 = 0.25, 0.25 \times 100 = 25\%$

Fractions to decimals

Convert simple fractions into decimals by division.

$\frac{5}{8} = 5 \div 8 = 0.625$

$$8\overline{)5.\,{}^50\,{}^20\,{}^40}$$
$$0.\,6\;2\;5$$

Use a written method if no calculator.

Another simple way of converting fractions into decimals is by finding an equivalent fraction with a denominator of 100.

$\frac{7}{20}$ is equivalent to $\frac{35}{100} = 0.35$

Multiply the numerator and denominator by 5.

$\frac{21}{25}$ is equivalent to $\frac{84}{100} = 0.84$

Multiply the numerator and denominator by 4.

Therefore $\frac{7}{20} = 0.35$ and $\frac{21}{25} = 0.84$

When ordering a mixture of fractions, decimals and percentages, convert all the values to the same type (e.g. all into decimals).

Write the following in order, from smallest to largest: $\frac{2}{5}$, $\frac{1}{4}$, 0.3, 28%

$\frac{2}{5} = 0.4$, $\frac{1}{4} = 0.25$ and 28% = 0.28

So $\frac{1}{4}$, 28%, 0.3, $\frac{2}{5}$ is the correct order.

Terminating decimals to fractions

A **terminating decimal** is a decimal number where the digits after the decimal point do not go on forever, e.g. 0.2, 0.35 and 0.875.

Convert decimals into fractions using $\frac{1}{10}$, $\frac{1}{100}$ and $\frac{1}{1000}$, etc.

$0.8 = \frac{8}{10} = \frac{4}{5}$

$0.48 = \frac{48}{100} = \frac{12}{25}$

Use HCF to cancel fractions to their simplest form.

$0.134 = \frac{134}{1000} = \frac{67}{500}$

Multiplying by a decimal

It is best to ignore the decimals and multiply the whole numbers using a written method.

Look at these examples.

4.2×3.8

	40	2	
	1200	60	30
	320	16	8

```
    1 2 0 0
      3 2 0
        6 0
  +     1 6
  ---------
    1 5 9 6
```

$42 \times 38 = 1596$

$4.2 \times 3.8 = 15.96$ ← A quick estimate will help place the decimal point ($4.2 \times 3.8 \approx 4 \times 4 = 16$)

2.23×4.9

	200	20	3	
	8000	800	120	40
	1800	180	27	9

```
        2 2 3
    ×     4 9
    ---------
      2 0 0 7
      8 9 2 0
    ---------
    1 0 9 2 7
```

$223 \times 49 = 10\,927$

$2.23 \times 4.9 = 10.927$ ← A quick estimate will help place the decimal point ($2.23 \times 4.9 \approx 2 \times 5 = 10$)

KEYWORDS

Terminating decimal ➤ A decimal number in which the digits do not repeat continuously.

Divisor ➤ The number which you are dividing by.

Adding and subtracting decimals

Use a column method to add or subtract decimals, e.g. $4.3 - 2.82$

```
    ³4 . ¹²3 ¹0      ← Place a zero in the
  -   2 . 8   2        hundredths column.
  -----------
      1 . 4   8
```

Dividing by a decimal

It is best to change the **divisor** to a whole number.

$45 \div 0.3 = 450 \div 3 = 150$ ← Multiplying both numbers by 10 gives the exact same calculation.

$1.23 \div 0.15 = 123 \div 15 = 8.2$ ← Multiply both numbers by 100.

Draw a 5 × 5 grid onto squared paper. Shade $\frac{7}{25}$ of the grid.

Draw an extra line in the centre of every row and column to turn it into a 10 × 10 grid.

Count the shaded squares to find out what $\frac{7}{25}$ is as a percentage and a decimal.

1. Convert the following fractions to decimals.
   (a) $\frac{3}{5}$ (b) $\frac{7}{8}$ (c) $\frac{2}{3}$ (d) $\frac{13}{20}$

2. Convert the following decimals to fractions.
   (a) 0.36   (b) 0.248   (c) 0.08

3. Without a calculator, work out:
   (a) 6.3×4.9 (b) $12 \div 0.4$ (c) $7.2 \div 0.5$

Rounding

When rounding you must always look at the digit in the next column:

If it is 5 or more, you round up.

Example	Nearest whole	Nearest ten
382.5	383	380
86.4	86	90
59.7	60	60

If it is 4 or less, you round down.

Example	Nearest hundred	Nearest thousand
6482	6500	6000
3549	3500	4000
10 859	10 900	11 000

Decimal places

It is important to give an answer to an appropriate **degree of accuracy**. Remember that you should never round a number until you have the final answer of a calculation.

Units	Usual degree of accuracy
Centimetres (cm)	1 decimal place
Metres (m)	2 decimal places
Kilometres (km)	3 decimal places

Rounding to two decimal places (2 d.p.) will leave you with two digits after the decimal point.

Round 3.476 32 to 2 d.p.

Round up since the next digit on the right is 5 or more.

$$3 . 4\,7|6\,3\,2 \;=\; 3 . 4\,8 \text{ (to 2 d.p.)}$$

Round after the second digit. Look at the next digit on the right.

Round 2.459 73 to 3 d.p.

Since the fourth digit is a 7 and the third digit is a 9, round the second digit up by 1.

$$2 . 4\,5\,9|7\,3 \;=\; 2 . 4\,6\,0 \text{ (to 3 d.p.)}$$

Round after the third digit.

Significant figures

Rounding to significant figures (s.f.) is useful for very large and very small numbers. The first significant figure of any number is the first digit which is not zero.

2 4 0 3 2 0   0 . 0 2 4 0 3 2 0

The next digits (4, 0, 3 etc.) on the right are the second, third, fourth (etc.) significant figures.

First s.f.

Example	1 s.f.	2 s.f.	3 s.f.
40 857	40 000	41 000	40 900
0.004 592	0.005	0.0046	0.004 59
12.93	10	13	12.9
1.987	2	2.0	1.99

KEYWORDS

Degree of accuracy ➤ What a number has been rounded to.

Upper bound ➤ The highest possible limit.

Lower bound ➤ The lowest possible limit.

Estimation

It is important to have an approximate idea of an answer before starting any calculation.
To estimate a calculation, it is best to round each number to 1 significant figure.

Estimate $\dfrac{3.4 \times 37}{0.21}$

$\dfrac{3.4 \times 37}{0.21} \approx \dfrac{3 \times 40}{0.2} = \dfrac{120}{0.2}$ ← Multiply the numerator and the denominator by 10.

$= \dfrac{1200}{2} = 600$

Estimate, to 1 significant figure, the total cost for 391 pupils to go on a school trip which costs £23 per person.

$391 \times 23 \approx 400 \times 20 = £8000$

Use a tape measure to find the length and width of your bedroom to the nearest centimetre.

What are the limits of accuracy for the length and width of your bedroom?

Limits of accuracy

When a measurement is taken, it is usually rounded to an appropriate degree of accuracy. The limits of accuracy are the maximum and minimum possible values for the measurement:

➤ The **upper bound** is the maximum possible value the measurement could have been.
➤ The **lower bound** is the minimum possible value the measurement could have been.

The bounds are exactly half the degree of accuracy above and below the measurement.

You can write the limits of accuracy using inequalities as shown.

Measurement	Degree of accuracy	Lower bound	Upper bound	Limits of accuracy
23mm	Nearest mm	22.5mm	23.5mm	$22.5 \leqslant x < 23.5$
54.6cm	1 decimal place	54.55cm	54.65cm	$54.55 \leqslant x < 54.65$
4.32m	2 decimal places	4.315m	4.325m	$4.315 \leqslant x < 4.325$
34000km	2 significant figures	33500km	34500km	$33500 \leqslant x < 34500$

Upper bound is actually 23.499...

We use the < symbol to mean 'up to but not including' 23.5

1. Round the following.
   (a) 38.4 to the nearest whole number
   (b) 365 to the nearest ten
   (c) 439 to the nearest hundred
   (d) 4505 to the nearest thousand
2. Round the following.
   (a) 2.3874 to 2 d.p.   (b) 4.60992 to 3 d.p.
   (c) 43549 to 2 s.f.   (d) 0.0040349 to 3 s.f.

3. Estimate $\dfrac{28 \times 4.9}{0.18}$
4. Write down, using inequalities, the limits of accuracy of the following measurements which have been rounded to the accuracy stated.
   (a) 23m (nearest m)   (b) 4.7kg (1 d.p.)
   (c) 5.23km (3 s.f.)

Calculator buttons

On a scientific calculator there are several buttons which are essential to know.

The table shows the button and its use. ← *Some calculators are different, so make sure you know how to use yours.*

Button	Use	Example
x^2	**Square**	$16^2 \longrightarrow$ [16] [x^2] [=] $\longrightarrow 256$
x^3	**Cube**	$7^3 \longrightarrow$ [7] [x^3] [=] $\longrightarrow 343$
y^x	Power	$3^6 \longrightarrow$ [3] [y^x] [6] [=] $\longrightarrow 729$
$\sqrt{\ }$	Square root	$\sqrt{60} \longrightarrow$ [$\sqrt{\ }$] [60] [=] $\longrightarrow 7.75$ (to 2 d.p.)
$\sqrt[3]{\ }$	Cube root	$\sqrt[3]{125} \longrightarrow$ [2nd F] [$\sqrt[3]{\ }$] [125] [=] $\longrightarrow 5$
$\sqrt[x]{\ }$	xth root	$\sqrt[5]{32} \longrightarrow$ [5] [2nd F] [$\sqrt[x]{\ }$] [32] [=] $\longrightarrow 2$
$a\frac{b}{c}$	Fraction	$\frac{8}{5} \longrightarrow$ [8] [$a\frac{b}{c}$] [5] [=] $\longrightarrow 1\frac{3}{5}$
$\times 10^x$	Standard form	$3.4 \times 10^4 \longrightarrow$ [3.4] [$\times 10^x$] [4] [=] $\longrightarrow 34\,000$

Some calculators use the button [ ]

Some calculators use the button [Exp]

Using brackets

Typing a large calculation into a calculator will usually require the use of brackets.

Calculate $\dfrac{23 + 4.5^2}{6.9 - 2.3}$

Type [(] [23] [+] [4.5] [x^2] [)] [÷] [(] [6.9] [−] [2.3] [)] [=] 9.402 173 913 ← *Do not round off unless directed to do so.*

It can be beneficial to do a check by calculating the numerator and denominator separately.

$\dfrac{23 + 4.5^2}{6.9 - 2.3} = \dfrac{43.25}{4.6}$ ← *Show this working in an exam.*

$= 9.402\,173\,913$

7

Module 7

Answers Using Technology

Standard form

Standard form calculations can be performed on a calculator. However, brackets are essential to separate the numbers.

Calculate $(2.1 \times 10^3) \times (4.5 \times 10^8)$

Type (2.1 ×10ˣ 3) × (4.5 ×10ˣ 8) = 9.45×10^{11} ← The calculator will display this answer in standard form.

Use the (−) button for negative indices.

Calculate $(3.6 \times 10^4) \div (2.5 \times 10^{-3})$

Type (3.6 ×10ˣ 4) ÷ (2.5 ×10ˣ (−) 3) = 14 400 000

$= 1.44 \times 10^7$ ← Change the answer back into standard form if required.

Make the number 16 by using the following calculator buttons.

| 3 | | 4 | | 5 | | (| |) | | $\sqrt[3]{}$ | | y^x | | + | | = |

You must only use each button once.

---

1. Calculate the following.
   (a) 2.3^3
   (b) 4^6
   (c) $\sqrt[3]{512}$

2. Calculate the following. Write down all the figures on your calculator display.

   $$\frac{8 + 2.5^2}{\sqrt{20.25}}$$

3. Calculate the following. Give your answer in standard form to 3 s.f.
   (a) $(2.6 \times 10^5) \times (3.8 \times 10^{-2})$
   (b) $(4.7 \times 10^8) \div (6.3 \times 10^{-7})$

Number Mind Map

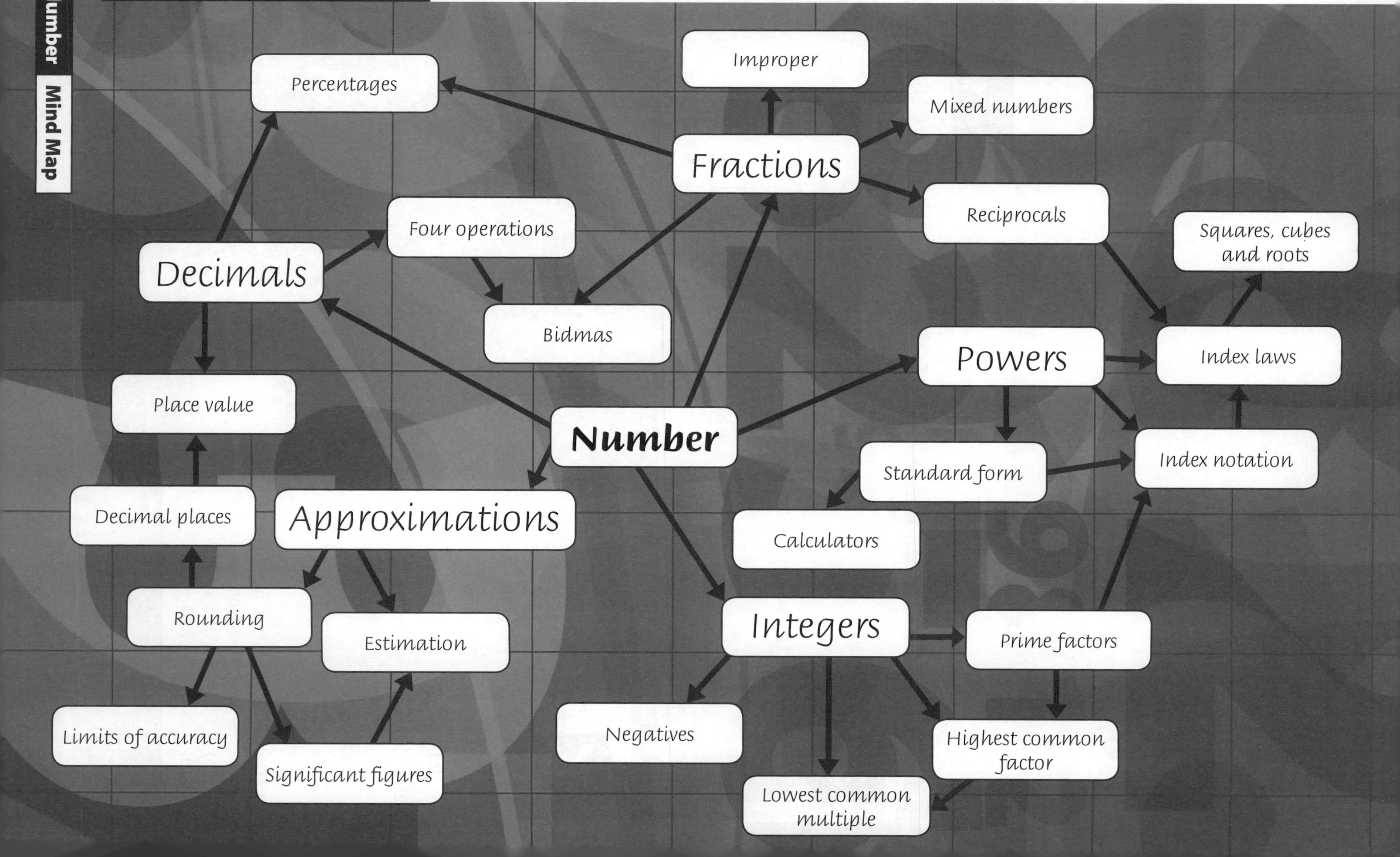
Number
Fractions
Improper
Mixed numbers
Reciprocals
Percentages
Decimals
Four operations
Bidmas
Place value
Powers
Index laws
Squares, cubes and roots
Index notation
Standard form
Calculators
Approximations
Decimal places
Rounding
Estimation
Limits of accuracy
Significant figures
Integers
Prime factors
Negatives
Highest common factor
Lowest common multiple

1. Work out the following. 📱
   (a) 4×5.8 **[2]**   (b) 2.3×42.7 **[3]**
   (c) $24 \div 0.08$ **[2]**   (d) $46.8 \div 3.6$ **[2]**
   (e) $-3 + -4$ **[1]**   (f) $-2 - -5 + -6$ **[1]**
   (g) $45 \div -9$ **[1]**   (h) $-4 \times -5 \times -7$ **[1]**
   (i) $4 + 3^2 \times 7$ **[1]**   (j) $(5 - 6^2) - (4 + \sqrt{25})$ **[2]**

2. (a) Write 36 as a product of its prime factors. Write your answer in index form. 📱 **[2]**
   (b) Write 48 as a product of its prime factors. Write your answer in index form. 📱 **[2]**
   (c) Find the highest common factor of 48 and 36. 📱 **[2]**
   (d) Find the lowest common multiple of 48 and 36. 📱 **[2]**

3. (a) Work out the following.
   $$\frac{4}{3} \times \frac{5}{6}$$
   Write your answer as a mixed number in its simplest form. 📱 **[2]**
   (b) The sum of three mixed numbers is $7\frac{11}{12}$. Two of the numbers are $2\frac{3}{4}$ and $3\frac{5}{6}$.
   Find the third number and give your answer in its simplest form. 📱 **[3]**
   (c) Calculate $3\frac{1}{5} \div 2\frac{1}{4}$
   Give your answer as a mixed number in its simplest form. 📱 **[3]**

4. Simplify the following, leaving your answer in index form. 📱
   (a) $5^3 \div 5^{-5}$ **[1]**   (b) $3^4 \times 3^8 \times 3$ **[1]**   (c) $\dfrac{7^2 \times 7^5}{7^3}$ **[2]**

5. Calculate the value of the following. 📱
   (a) 2.3^0 **[1]**   (b) 5^4 **[1]**   (c) 3^{-2} **[2]**

6. (a) Write the following numbers in standard form. 📱
   (i) $67\,000\,000$ **[1]**   (ii) $0.000\,007$ **[1]**
   (iii) $43\,600$ **[1]**   (iv) $0.008\,03$ **[1]**
   (b) Calculate the following, giving your answers in standard form. 📱
   (i) $(4 \times 10^6) \times (3 \times 10^5)$ **[2]**
   (ii) $(2.1 \times 10^{-2}) + (5.7 \times 10^{-3})$ **[2]**
   (iii) $(1.2 \times 10^8) \div (3 \times 10^4)$ **[2]**

7. Paul's garage measures 6m in length to the nearest metre. His new car measures 5.5m in length to 1 decimal place.
   Write down, using inequalities, the limits of accuracy of:
   (a) Paul's garage **[2]**
   (b) Paul's new car **[2]**

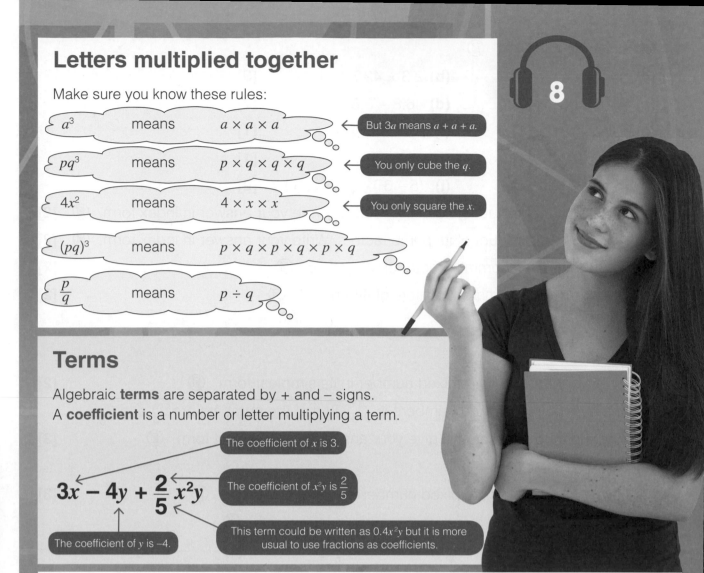

Letters multiplied together

Make sure you know these rules:

a^3 means $a \times a \times a$ ← But $3a$ means $a + a + a$.

pq^3 means $p \times q \times q \times q$ ← You only cube the q.

$4x^2$ means $4 \times x \times x$ ← You only square the x.

$(pq)^3$ means $p \times q \times p \times q \times p \times q$

$\frac{p}{q}$ means $p \div q$

Terms

Algebraic **terms** are separated by + and – signs.
A **coefficient** is a number or letter multiplying a term.

The coefficient of x is 3.

The coefficient of x^2y is $\frac{2}{5}$

$$3x - 4y + \frac{2}{5}x^2y$$

The coefficient of y is –4.

This term could be written as $0.4x^2y$ but it is more usual to use fractions as coefficients.

Simplifying expressions

An **expression** is a collection of (one or more) algebraic terms. 'Like terms' have the same combination of letters. Here is an example of an expression:

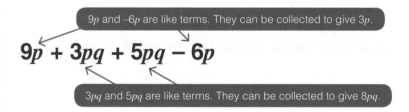

$9p$ and $-6p$ are like terms. They can be collected to give $3p$.

$$9p + 3pq + 5pq - 6p$$

$3pq$ and $5pq$ are like terms. They can be collected to give $8pq$.

So $9p + 3pq + 5pq - 6p = 3p + 8pq$

Combinations of letters that are **not** the same are said to be **unlike** terms.
For example, $2xy$ and xy^2 are unlike terms. They can't be added together.

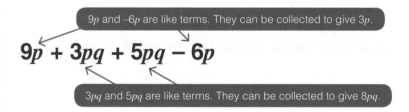

Term ➤ A single number or variable, or numbers and variables multiplied together.

Coefficient ➤ The number occurring at the start of each term.

Expression ➤ A collection of terms.

Indices (powers) ➤ A convenient way of writing repetitive multiplication, e.g. $3^4 = 3 \times 3 \times 3 \times 3 = 81$. In this example, the power of 3 is 4.

Identity ➤ An equation that is true for all values.

KEYWORDS

Module 8   Algebraic Notation

Laws of indices

Make sure you know these seven rules:

Rule in words	Rule in symbols	Examples
When multiplying numbers with the same base, you add the **indices**	$a^m \times a^n = a^{m+n}$	$p^4 \times p^3 = p^{(4+3)} = p^7$ $2^7 \times 2^3 = 2^{10}$
When dividing numbers with the same base, you subtract the indices	$a^m \div a^n = a^{m-n}$	$q^8 \div q^3 = q^{(8-3)} = q^5$ $3^{13} \div 3^8 = 3^5$
When finding a power of a power, you multiply the powers	$(a^m)^n = a^{mn}$	$(a^2)^5 = a^{10}$ $(4^3)^4 = 4^{12}$
Any non-zero number to the power zero is 1	$a^0 = 1$	$q^0 = 1$ $8^0 = 1$
Negative powers can be made positive by using the reciprocal	$a^{-n} = \dfrac{1}{a^n}$	$a^{-3} = \dfrac{1}{a^3}$ $2^{-4} = \dfrac{1}{2^4} = \dfrac{1}{16}$
Fractional powers mean roots	$a^{\frac{1}{n}} = \sqrt[n]{a}$	$a^{\frac{1}{2}}$ means the square root of a, so $25^{\frac{1}{2}} = 5$ $a^{\frac{1}{3}}$ means the cube root of a, so $64^{\frac{1}{3}} = 4$
Fractional powers may be split into finding a root, followed by finding a power*	$a^{\frac{m}{n}} = \left(a^{\frac{1}{n}}\right)^m$	$125^{\frac{2}{3}} = \left(125^{\frac{1}{3}}\right)^2 = 5^2 = 25$

*Also remember – when finding the power of a **fraction**, you find the power of both the numerator and denominator of the fraction. So $\left(\dfrac{8}{27}\right)^{\frac{1}{3}} = \dfrac{8^{\frac{1}{3}}}{27^{\frac{1}{3}}} = \dfrac{\sqrt[3]{8}}{\sqrt[3]{27}} = \dfrac{2}{3}$

Equations and identities

An equation can be solved to find an unknown quantity.

$5x + 3 = 23$ can be solved to find $x = 4$.

An **identity** is true for all values of x. $(x + 3)^2$ is identically equal to $x^2 + 6x + 9$.

You can write this as $(x + 3)^2 \equiv x^2 + 6x + 9$

Make nine cards with the letters $a, a, a, b, b, b, c, c, c$ on them. Turn all the cards face down and mix them up. Turn over two cards and multiply the letters, writing the answer algebraically, e.g. ab. Then repeat by picking three cards and multiplying the letters.

1. Simplify
    (a) $3ab - 2a^2b + 9ab + 4a^2b$
    (b) $2x - 7 + 7x - 2$
    (c) $ab + bc + ca + ac + cb + ba$

2. Simplify
    (a) $g^{12} \div g^4$
    (b) $\left(p^{15}\right)^{\frac{1}{3}}$
    (c) $2p^4q^{-3} \times 7p^{-2}q$

3. Evaluate
    (a) $8^{\frac{2}{3}}$
    (b) 4^{-2}
    (c) $\left(\dfrac{2}{3}\right)^{-3}$

Brackets

You need to be able to work with brackets:

Rule	Examples
To **multiply out** single brackets, multiply the term outside the brackets by every term inside.	$5(2a - 6) = (5 \times 2a) - (5 \times 6) = 10a - 30$ $3p(6p + 2q) = (3p \times 6p) + (3p \times 2q) = 18p^2 + 6pq$
To multiply out double brackets, multiply every term in the first bracket by every term in the second bracket.	$(2x + 3)(x - 6) = (2x \times x) + (2x \times -6) + (3 \times x) + (3 \times -6)$ $= 2x^2 - 12x + 3x - 18$ $= 2x^2 - 9x - 18$
When squaring brackets, always write out the expression in full to avoid making mistakes.	$(3x - 2)^2 = (3x - 2)(3x - 2) = 9x^2 - 6x - 6x + 4$ $= 9x^2 - 12x + 4$ $(3x - 2)^2$ is **not** equal to $9x^2 + 4$

Double brackets can also be multiplied out using a 'box method'.
For example, if you wanted to expand $(2x + 1)(3x - 2)$

×	**2x**	**+1**
3x	$6x^2$	$3x$
−2	$-4x$	-2

So $(2x + 1)(3x - 2) = 6x^2 + 3x - 4x - 2$
$= 6x^2 - x - 2$

Square brackets may also be multiplied out using this method.
To expand $(3 - 2p)^2$

×	**3**	**−2p**
3	9	$-6p$
−2p	$-6p$	$4p^2$

So $(3 - 2p)^2 = 9 - 6p - 6p + 4p^2$
$= 9 - 12p + 4p^2$

Factorising

Factorising is the reverse of multiplying out brackets. You have to look for the common factors in every term of the expression.

$12p + 15q = 3(4p + 5q)$ ← First look at the numbers. 12 and 15 have a highest common factor of 3. The letters are p and q, and these do not appear in both terms, so you can only 'factor out' a 3.

$6xy + 16py = 2y(3x + 8p)$ ← 6 and 16 have a highest common factor of 2. The letter y appears in both terms. So the common factor here is $2y$.

If the expression includes powers, look for the largest number and highest power that goes into each term.

$10x^2 - 5x = 5x(2x - 1)$ ← $5x$ is the largest factor of $10x^2$ and $5x$.

$6x^5 - 4x^2 = 2x^2(3x^3 - 2)$ ← 2 is the largest factor of both 6 and 4. x^2 is the highest power of x that goes into x^5 and x^2.

$18a^2b^6 - 12ab^4 = 6ab^4(3ab^2 - 2)$ ← $6ab^4$ is the largest factor of $18a^2b^6$ and $12ab^4$.

A **quadratic expression** is an expression of the form $ax^2 + bx + c$.

Algebraic Expressions

Module 9

Factorising expressions

$x^2 + 5x - 24$ ← Look for two numbers that multiply to give –24 and add to give 5. In this case and $8 \times (-3) = -24$ and $8 + (-3) = 5$

So $x^2 + 5x - 24 = (x + 8)(x - 3)$

$x^2 + 18x + 65$ ← Look for two numbers that multiply to give 65 and add to give 18. The numbers required are 13 and 5.

So $x^2 + 18x + 65 = (x + 13)(x + 5)$

Always check you have the right answer by expanding the brackets: $(x + 13)(x + 5)$ $= x^2 + 13x + 5x + 65 = x^2 + 18x + 65$

Difference of two squares

Expressions with one square subtracted from another square can be factorised. These examples all follow the pattern $a^2 - b^2 \equiv (a + b)(a - b)$, an identity known as the **difference of two squares**.

$a^2 - 9 = a^2 - 3^2 = (a + 3)(a - 3)$

$16x^2 - 25y^6 = (4x)^2 - (5y^3)^2 = (4x + 5y^3)(4x - 5y^3)$

$3a^2 - 3 = 3(a^2 - 1) = 3(a + 1)(a - 1)$

Write these six expanded expressions on separate pieces of paper and mix them up:

$x^2 + 6x + 9$	$x^2 - 9$	$x^2 - 6x - 9$
$x^2 - 6x + 9$	$x^2 + 6x - 27$	$x^2 - 9x$

Write these five factorised expressions on separate pieces of paper and mix them up:

$(x + 3)(x - 9)$ $x(x - 9)$ $(x + 3)^2$ $(x + 3)(x - 3)$ $(x - 3)^2$

Try to pair each expanded expression with the correct factorised expression. One of the expanded expressions cannot be factorised.

1. Multiply out
   (a) $5(2 - 3x)$
   (b) $(x + 5)(x - 7)$

2. Factorise
   (a) $x^2 - x - 56$
   (b) $4p^2 - 25q^2$

3. Multiply out
   (a) $(2x + 3)(x - 1)$
   (b) $(3x + y)(3x - y)$

KEYWORDS

Multiplying out ➤ Removing brackets from an algebraic expression by multiplication.

Factorising ➤ Writing an algebraic expression as a product of factors, usually in brackets.

Difference of two squares ➤ The algebraic identity $a^2 - b^2 \equiv (a + b)(a - b)$

Using and rearranging formulae

Rearranging formulae is very similar to solving equations.

Make x the subject in the equation
$$2x - y = z$$

$2x = z + y$ ← Put all terms excluding x on the other side of the equation.

$x = \dfrac{z + y}{2}$ ← Divide through by 2.

Make x the subject in the equation
$$\frac{2 - 3x}{4x + 1} = y$$
← If the subject appears twice in the equation, you are likely to need to factorise once all the required terms are on one side of the equation.

$2 - 3x = y(4x + 1)$ ← Clear the fraction.

$2 - 3x = 4xy + y$ ← Multiply out the brackets.

$2 = 3x + 4xy + y$ ← Collect all x terms on one side.

$2 - y = 3x + 4xy$ ← Put all terms not including x on one side.

$2 - y = x(3 + 4y)$ ← Factorise the terms involving x.

$x = \dfrac{2 - y}{3 + 4y}$ ← Divide through by $3 + 4y$.

Make x the subject in the equation
$$\frac{5 + 12y}{x + 2y} = 8$$

$5 + 12y = 8(x + 2y)$ ← Multiply through by $x + 2y$ to clear the fraction.

$5 + 12y = 8x + 16y$ ← Multiply out the brackets.

$5 - 4y = 8x$ ← Put all terms not including x on one side.

$x = \dfrac{5 - 4y}{8}$ ← Divide through by 8.

Make x the subject in the equation
$$p + q = \sqrt{10x + y}$$

$(p + q)^2 = 10x + y$ ← Square both sides to clear the square root sign.

$(p + q)^2 - y = 10x$ ← Put all terms not including x on one side.

$\dfrac{(p + q)^2 - y}{10} = x$ ← Divide through by 10.

Substituting numerical values in formulae

It is important to **substitute** numbers accurately into a given formula.

10

If $C = 3r^2$, find C when $r = 5$
$C = 3 \times 5^2 = 3 \times 25 = 75$ ← C is **not** $(3 \times 5)^2$. You only square the 5.

If $v = u + at$, find v when $u = 10$, $a = -2$ and $t = 8$
$v = 10 + (-2) \times 8 = 10 - 16 = -6$

If $v^2 = u^2 + 2as$, find v when $u = 5$, $a = -3$ and $s = 2$
$v^2 = 5^2 + 2 \times -3 \times 2 = 25 - 12 = 13$
So $v = \sqrt{13} = 3.61$ (2 d.p.)

If $\dfrac{1}{f} = \dfrac{1}{u} + \dfrac{1}{v}$, find v when $f = 5$ and $u = 12$
$\dfrac{1}{5} = \dfrac{1}{12} + \dfrac{1}{v}$, so $\dfrac{1}{v} = \dfrac{1}{5} - \dfrac{1}{12}$
$\dfrac{1}{v} = \dfrac{12}{60} - \dfrac{5}{60} = \dfrac{7}{60}$
So $v = \dfrac{60}{7} = 8.57$ (2 d.p.)

← You could rearrange this formula to make v the subject first, though it is usually easier to substitute the numbers straight in.

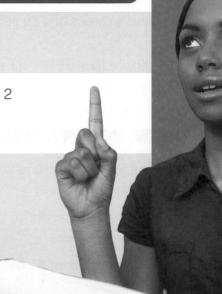

Algebraic Formulae

Module 10

Translating procedures into expressions or formulae

You may be asked to derive a formula from some given information.

A motorist wishes to rent a car. The rental cost is £35 per day, plus a standing charge of £60. Write down an expression for the total cost, £C, in terms of the number of days, d. For how many days can the motorist rent the car if he has £400 to spend?

The total cost is $C = 60 + 35d$

If $C = 400$, then you need to solve the equation $400 = 60 + 35d$

So $340 = 35d$ and $d = \dfrac{340}{35} = 9.71$

The motorist can rent the car for a maximum of 9 days.

The angles in a quadrilateral, in degrees, are $6y + 16$, $24 + y$, $106 - y$ and 130. Form an equation and use it to find the value of y.

The interior angles of a quadrilateral add up to $360°$, so $6y + 16 + 24 + y + 106 - y + 130 = 360$

Therefore $6y + 276 = 360$

So $6y = 84$ and $y = \dfrac{84}{6} = 14°$

S is the total sum of the interior angles in a regular polygon. If the polygon has n sides, find an expression for S in terms of n.

If the polygon has n sides, then it can be split up into $n - 2$ triangles.

For example, a regular six-sided hexagon can be split into four triangles.

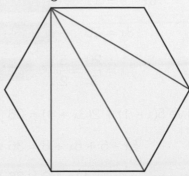

The interior angles of each triangle add up to $180°$, therefore $S = 180(n - 2)$

1. Make p the subject in each of the following formulae.

   (a) $3p + a = b$ (b) $q = \dfrac{3p + 2}{2p + 3}$

2. If $a = -10$, $b = 20$ and $c = 4$, find the exact values for p when

   (a) $p = a^2 - c^2$ (b) $p = \dfrac{a + c}{b}$

3. An equilateral triangle has interior angles of size $3x$, $x - 2$ and $2x - 10$.
   (a) Write down an equation in x.
   (b) Solve the equation to find x.
   (c) Find the values of the three angles.

The area of a trapezium is given by $A = \dfrac{(a + b)h}{2}$

Write each letter and symbol in this equation on separate pieces of card. You will also need a card with the minus symbol on it. Rearrange the letters and symbols to make a the subject of the equation. Then do the same to make b the subject and finally to make h the subject.

Solving linear equations

To solve any **linear equation**, you rearrange it to make x the subject.

> Remember to do the same operation to both sides of the equation.

Solve $3x + 4 = 25$

$3x = 21$ ← Subtract 4 from both sides.

$x = 7$ ← Divide both sides by 3.

Solve $12x - 9 = 4x + 11$

$8x - 9 = 11$ ← Subtract $4x$ from both sides.

$8x = 20$ ← Add 9 to both sides.

$x = \dfrac{20}{8} = \dfrac{5}{2}$ ← Divide both sides by 8.

Solve $5(x - 1) + 2(3x + 4) = 36$

$5x - 5 + 6x + 8 = 36$ ← Multiply out the brackets.

$11x + 3 = 36$ ← Collect 'like terms'.

$11x = 33$ ← Subtract 3 from both sides.

$x = \dfrac{33}{11}$ ← Divide both sides by 11.

$x = 3$

Choose a positive whole number b.

Write down the quadratic equation $x^2 - 2bx + b^2 = 0$ using your value of b.

Try to solve your equation. What do you notice?

Now try this again with a different value for b. Does the same thing happen again? Can you see why?

Solving quadratic equations

Quadratic equations are usually written in the form $ax^2 + bx + c = 0$, where a, b and c are numbers.

Solving by factorising ←

> Factorising uses the idea that if two expressions multiplied together gives 0, then one of the expressions must be equal to 0.

Solve $x^2 + 11x + 18 = 0$

$(x + 9)(x + 2) = 0$ ← Factorise the left-hand side.

So $x + 9 = 0$ or $x + 2 = 0$

Therefore $x = -9$ or $x = -2$

Solve $x^2 - 12x + 35 = 0$

$(x - 5)(x - 7) = 0$

So $x - 5 = 0$ or $x - 7 = 0$

Therefore $x = 5$ or $x = 7$

Solve $x^2 + 7x - 18 = 0$

$(x + 9)(x - 2) = 0$

So $x + 9 = 0$ or $x - 2 = 0$

Therefore $x = -9$ or $x = 2$

Solve $x^2 - 64 = 0$ ←

> Watch out for the 'difference of two squares'.

$(x + 8)(x - 8) = 0$ ← Factorise.

So $x + 8 = 0$ or $x - 8 = 0$

Therefore $x = -8$ or $x = 8$

Simultaneous linear equations

You can solve **simultaneous equations** by **elimination**.

Solve the simultaneous equations
$$3x + 5y = 23$$
$$3x + 2y = 11$$

$3y = 12$ ← Now divide both sides by 3.

The coefficients of x and y are the same (equal to 3) so **subtract** both equations to eliminate x.

$y = 4$ ← Now substitute $y = 4$ back into either of the original equations. If you choose the second equation: $3x + 8 = 11$ so $3x = 3$ and $x = 1$

The solution is therefore $x = 1$, $y = 4$

To check these values are correct, you can substitute them into either of the initial equations:

$3 \times 1 + 5 \times 4 = 23$ $3 \times 1 + 2 \times 4 = 11$

Solve the simultaneous equations

$2x - 5y = 23$

$x + y = 1$ ← The coefficients of x or those of y must be the same to **eliminate** them. Here neither are the same but the 2nd equation can be multiplied by 2.

$2x - 5y = 23$

$2x + 2y = 2$ ← Now the coefficients of x are both 2, so **subtract** to eliminate x.

$-7y = 21$, so $y = -3$

$x + y = 1$ ← Substitute $y = -3$ back into either of the original equations.

$x - 3 = 1$, so $x = 4$

So the solutions are $x = 4$, $y = -3$

To check these are correct:

$2 \times 4 - 5 \times (-3) = 23$ $4 + (-3) = 1$

Callie regularly visits the sweet shop. One week she bought five chocolate bars and three bags of fruit drops, costing £5.95. The next week she bought two chocolate bars and one bag of fruit drops, costing £2.25. Find the cost of one chocolate bar and one bag of fruit drops.

Let c = the cost of one chocolate bar

Let d = the cost of one bag of fruit drops

Then $5c + 3d = 595$

And $2c + d = 225$ ← Set up two equations using the information given in the question.

Rewrite as:

$6c + 3d = 675$ ← Multiplying the second equation by 3.

And $5c + 3d = 595$ ← Leave the first equation as it is. You now have $3d$ in each equation.

So $c = 80$ ← Subtracting these two equations gives the cost of one chocolate bar (80p).

Finally, $2 \times 80 + d = 225$, or $160 + d = 225$

So $d = 65$ ← 65p is the cost of one bag of fruit drops.

Using graphs to solve quadratic equations

Graphs can be used to solve equations by looking at their points of intersections with the x-axis or with each other.

This graph shows the equation
$y = x^2 + 5x + 3$
Use it to solve the equation $x^2 + 5x + 3 = 0$

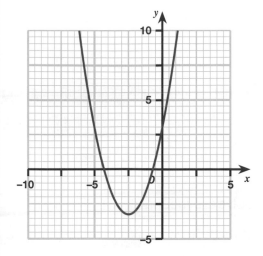

$x^2 + 5x + 3 = 0$ is where the curve crosses the x-axis, so the solutions are (approx.) $x = -4.3$ and $x = -0.8$

By drawing a straight line on the same grid you can solve $x^2 + 4x - 2 = 0$

Start with the equation **you are trying to solve**, and add or subtract x^2 terms, x terms and numbers to obtain the equation **you are given**.

Adding $x + 5$ to both sides of $x^2 + 4x - 2 = 0$ gives $x^2 + 5x + 3 = x + 5$

You already have the graph of $y = x^2 + 5x + 3$, so plot $y = x + 5$ and find the x-coordinates of the point(s) of intersection. You will find the solutions are (approx.) $x = -4.5$ and $x = 0.5$

1. Solve these equations.
    (a) $7(x - 2) = 35$
    (b) $4 - 2x = 5 - 5x$

2. Solve the equation $x^2 - 11x + 10 = 0$

3. Solve the simultaneous equations
    $$4x - y = 13$$
    $$3x + y = 1$$

4. Solve the equation $\dfrac{x + 1}{2} + \dfrac{x - 4}{3} = \dfrac{10}{3}$

Inequality symbols

The **inequality** symbols are:

< means less than	≤ means less than or equal to
> means greater than	≥ means greater than or equal to

Solving linear inequalities in one variable

Linear inequalities are solved in the same way as linear equations, with one main exception: when multiplying or dividing by a negative number, you change the direction of the inequality symbol.

Solve the inequality $5x + 4 \geq 2x - 20$

$3x + 4 \geq -20$ ← Subtract $2x$ from both sides.

$3x \geq -24$ ← Subtract 4 from both sides.

$x \geq -8$ ← Divide by 3.

Solve the inequality $x + 33 > 4x + 24$

$33 > 3x + 24$ ← Subtract x from both sides.

$9 > 3x$ ← Subtract 24 from both sides.

$3 > x$ ← Divide both sides by 3.

$x < 3$ ← Rewrite the inequality starting with x (3 is greater than x means that x is less than 3).

Solve the inequality $12 - 5x > 7$

$-5x > -5$ ← Subtract 12 from both sides.

$x < 1$ ← Both sides are divided by a negative number (−5) so the direction of the inequality is changed.

Each of these solutions may be represented on a number line.

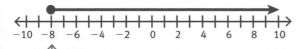

$x \geq -8$ is represented by a 'filled-in' circle (since −8 is included) and an arrow to the right.

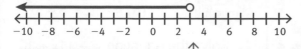

$x < 3$ is represented by a 'hollow' circle (since 3 is excluded) and an arrow to the left.

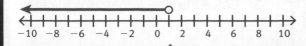

$x < 1$ is represented by a 'hollow' circle and an arrow to the left.

Module 12

Algebraic Inequalities

Inequalities involving fractions

If the inequality involves fractions, you should 'multiply through' to remove the fractions.

Solve the inequality $\dfrac{5x - 9}{3} > 2$

$5x - 9 > 6$ ← Multiply both sides by 3.

$5x > 15$ ← Add 9 to both sides.

$x > 3$ ← Divide both sides by 5.

Two inequalities

If there are two inequalities, then divide the inequality into two parts and solve each part separately.

Solve $5 < 4x - 3 \leqslant 21$

Split this into $5 < 4x - 3$ and $4x - 3 \leqslant 21$

$5 < 4x - 3$	$4x - 3 \leqslant 21$
$8 < 4x$	$4x \leqslant 24$
$2 < x$	$x \leqslant 6$

So the solution can be written as $2 < x \leqslant 6$

The integer values that satisfy this inequality are $x = 3$, $x = 4$, $x = 5$ and $x = 6$. On a number line:

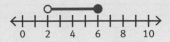

Solve $1 < \dfrac{3x - 1}{2} < 16$

Split this into $1 < \dfrac{3x - 1}{2}$ and $\dfrac{3x - 1}{2} < 16$

$2 < 3x - 1$	$3x - 1 < 32$
$3 < 3x$	$3x < 33$
$1 < x$	$x < 11$

So the solution is $1 < x < 11$

KEYWORDS

Inequality ➤ A relationship when one quantity is greater or less than another.

Linear inequality ➤ An inequality involving only x terms and numbers.

Chalk a number line from –5 to 5 on the ground. Practise showing these solutions to inequalities using filled-in circles, hollow circles and arrows:

$x \geqslant 1$ $x < 2$ $x > -4$ $x \leqslant 3$

1. Solve the inequalities:
   (a) $4x - 9 < x + 15$
   (b) $8 - 9x \geqslant 33 + x$

2. Write down the inequalities shown by these number lines:

   (a)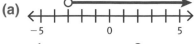

   (b)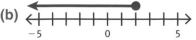

3. Solve the inequality $2(x - 6) < 5(4 - 2x)$ and show your solution on a number line.

Term-to-term and position-to-term sequences

A term-to-term **sequence** means you can find a rule for each **term** based on the previous term(s) in the sequence.

$$U_{n+1} = U_n + 4, \ U_1 = 3$$

> This means the next term in the sequence is found by adding 4 to the previous term, and the first term is 3.

> U_n just means 'the nth term in the sequence'.

The sequence is 3, 7, 11, 15, 19, 23, …

In a position-to-term sequence, you can find a rule for each term based on its position in the sequence.

$$U_n = 8n - 2$$

> Remember, U_n means 'the nth term in the sequence'.

> To find the first term, you substitute $n = 1$. To find the second term, you substitute $n = 2$, and so on.

The sequence is 6, 14, 22, 30, 38, …

Consider the term-to-term sequence $U_{n+1} = 3U_n$ with $U_1 = 2$

> In this sequence, the first term is 2 and you multiply by 3 to go from one term to the next.

First term in the sequence $\longrightarrow$ $U_1 = 2$

Second term in the sequence $\longrightarrow$ $U_2 = 3U_1 = 3 \times 2 = 6$

Third term in the sequence $\longrightarrow$ $U_3 = 3U_2 = 3 \times 6 = 18$

Fourth term in the sequence $\longrightarrow$ $U_4 = 3U_3 = 3 \times 18 = 54$

The sequence is 2, 6, 18, 54, 162, …

> The next term will be $162 \times 3 = 486$

This sequence may also be expressed using a position-to-term formula.

The formula is $U_n = 2 \times 3^{n-1}$. In other words, the nth term in the sequence is $2 \times 3^{n-1}$.

$$U_1 = 2 \times 3^{1-1} = 2 \qquad U_2 = 2 \times 3^{2-1} = 6 \qquad U_3 = 2 \times 3^{3-1} = 18 \qquad U_4 = 2 \times 3^{4-1} = 54$$

Standard sequences

Square numbers	$U_n = n^2$	
1, 4, 9, 16, 25, …	This sequence is produced by squaring the numbers 1, 2, 3, 4, 5, … etc.	
Cube numbers	$U_n = n^3$	
1, 8, 27, 64, 125, …	This sequence is produced by cubing the numbers 1, 2, 3, 4, 5, … etc.	
Triangular numbers	$U_n = \dfrac{n(n+1)}{2}$	
1, 3, 6, 10, 15, 21, …	This sequence is generated from a pattern of dots that form a triangle.	
Fibonacci sequence	$U_{n+2} = U_{n+1} + U_n$ and $U_1 = U_2 = 1$	There are other 'types' of Fibonacci sequence. One is the Lucas series: 2, 1, 3, 4, 7, 11, 18, … The term-to-term formula is the same as the original Fibonacci sequence, i.e. $U_{n+2} = U_{n+1} + U_n$, though in this case $U_1 = 2$ and $U_2 = 1$
1, 1, 2, 3, 5, 8, 13, 21, …	The next number is found by adding the two previous numbers. So the next number in the sequence would be $13 + 21 = 34$.	

The Fibonacci sequence is also a type of **recursive sequence** as finding the next term requires you to know previous terms.

Arithmetic sequences

An **arithmetic sequence** is a sequence of numbers having a common first difference.

Sequence   4 7 10 13 16 ← Here there is a common first difference of 3.

For any arithmetic sequence, the position-to-term formula is given by $U_n = dn + (a - d)$, where a is the first term and d is the common difference.

So here $U_n = 3n + (4 - 3)$, i.e. $U_n = 3n + 1$ ← The next number in the sequence will therefore be $U_6 = 3 \times 6 + 1 = 19$

Geometric sequences

In a **geometric sequence** you multiply by a constant number (common ratio) r to go from one term to the next.

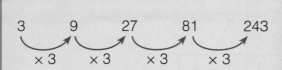

Each number is multiplied by 3 each time, so the common ratio $r = 3$

The term-to-term formula is $U_{n+1} = 3U_n$ with $U_1 = 3$

The position-to-term formula is $U_n = 3^n$

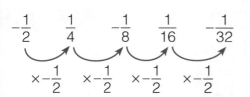

The common ratio $r = -\frac{1}{2}$

The term-to-term formula is $U_{n+1} = -\frac{U_n}{2}$ with $U_1 = -\frac{1}{2}$

The position-to-term formula is $U_n = \left(-\frac{1}{2}\right)^n$

One problem that uses sequences is called the 'handshake problem'.

Draw three people in a triangle.

If every person shakes hand with each other, how many handshakes will there?

Now draw four people in a square. Again, how many handshakes will there be?

If n is the number of people, it can be shown that the position-to-term formula for the number of handshakes is $U_n = \frac{n^2}{2} - \frac{n}{2}$

How many handshakes would take place if there were six people? Check by drawing a diagram **and** using the formula.

13

1. Find the first three terms in the sequence given by
   $$U_n = \frac{n^3 - 1}{2}$$

2. The sequence 4, 13, 22, 31, 40, … forms an arithmetic sequence.
   (a) Find an expression for U_n, the position-to-term formula.
   (b) Now find an exact value for U_{100}, the 100th term in the sequence.
   (c) Is 779 a number in this sequence? Explain your answer.

KEYWORDS

Sequence ➤ A series of numbers following a mathematical pattern.

Term ➤ A number in a sequence.

Arithmetic sequence ➤ A sequence with a common first difference between consecutive terms.

Geometric sequence ➤ A sequence with a common ratio.

Coordinates

Coordinates are used to describe any point in two-dimensional space. They are written using brackets (x, y). The x-coordinate tells you how far to the **right** the coordinate is, and the y-coordinate tells you how far **up**.

The x-axis and y-axis split the space into four quadrants. Here are examples of coordinates in each quadrant.

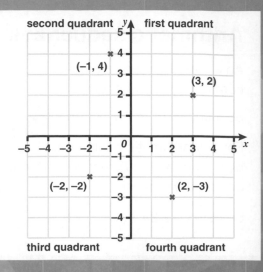

Graphs of straight lines

To plot a straight-line graph, you should first draw up a table of values.

14

Plot the graph of $y = -2x + 4$

x	0	1	2	3
y	4	2	0	-2

Plot the points and draw the line.

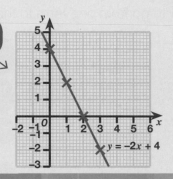

Take some simple values for x, perhaps 0, 1, 2 and 3, and calculate y for each value. For example, when $x = 3$, $y = -2 \times 3 + 4 = -2$

$y = mx + c$

The equation of any straight line can be written as $y = mx + c$ where m is the **gradient** and c is the point of intercept with the y-axis.

In the example above, the gradient m is -2 and the y-intercept, c, is 4. So the equation is $y = -2x + 4$

You need to be able to calculate both the gradient and the y-intercept from a given graph.

Find the equation of this straight line.

To find the gradient, take any two points on the line. Here we have chosen $(0, 3)$ and $(4, 5)$.

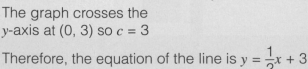

$$m = \frac{\text{change in } y}{\text{change in } x} = \frac{2}{4} = \frac{1}{2}$$

The graph crosses the y-axis at $(0, 3)$ so $c = 3$

Therefore, the equation of the line is $y = \frac{1}{2}x + 3$

Gradient ➤ The gradient, or slope, of a straight line shows how steep it is.

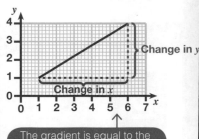

The gradient is equal to the $\frac{\text{change in } y}{\text{change in } x}$

Coordinates ➤ A pair of numbers (in brackets) used to determine a point in a plane.

Parallel ➤ Two lines are parallel if they have the same gradient. Parallel lines can never meet.

KEYWORDS

Module 14 Coordinates and Linear Functions

Finding the equation of a line through two given points

Here is a line going through the points (−2, 5) and (4, 1).

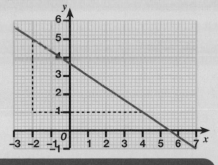

The gradient

$$m = \frac{\text{change in } y}{\text{change in } x} = \frac{-4}{6} = -\frac{2}{3}$$

Now the equation of the line is $y = mx + c$,

or $y = -\frac{2}{3}x + c$,

> Notice from (−2, 5) to (4, 1), the y value decreases, so the change in y is negative. The gradient is negative since the line slopes downwards.

Substituting $x = 4$, $y = 1$ gives $1 = -\frac{8}{3} + c$,

so $c = \frac{11}{3}$

> This can be written as $3y = -2x + 11$

The equation of the line is therefore $y = -\frac{2}{3}x + \frac{11}{3}$

Chalk two sets of x and y axes on the ground. On one set of axes, chalk a straight line going diagonally 'uphill'. On the second set of axes, chalk a straight line going diagonally 'downhill'. Pace out the changes in x and y to work out the rough gradient of each diagonal line.

Finding the equation of a line through one point with a given gradient

Find the equation of the line with gradient 3, going through (2, 4).

The equation of the line is $y = mx + c$, or $y = 3x + c$.

$4 = 3 \times 2 + c$

> Now substitute the **coordinates** of the point to find c, i.e. use $x = 2$ and $y = 4$

$4 = 6 + c$

So $c = -2$ and the equation of the line is $y = 3x - 2$

Parallel lines

Two lines are **parallel** if they have the same gradient. For example, $y = 2x + 1$, $y = 2x$ and $y = 2x - 3$ are all parallel.

A line L is parallel to the line $y = 2x + 4$ and passes through the point (2, 7). Find the equation of L.

Since both lines are parallel, they both have the same gradient, and so L has gradient $m = 2$

So L has equation $y = 2x + c$

Substituting $x = 2$ and $y = 7$ gives $7 = 2 \times 2 + c$, or $7 = 4 + c$

So $c = 3$ and L has equation $y = 2x + 3$

1. A line has gradient 3 and goes through the point (0, −2). Write down the equation of the line.
2. Three lines are given by $4y = 8 - x$, $y = 2x + 16$ and $4y = 8x - 1$. Which two lines are parallel?
3. A line has gradient 4 and goes through the point (8, 0). Find the equation of the line.
4. Find the equation of the line joining points (−2, −6) to (6, 12).

Roots and intercepts

A quadratic function has the equation
$y = ax^2 + bx + c$

If $a > 0$, the graph will will look like:

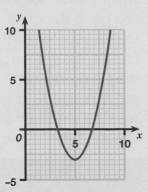

If $a < 0$, the graph will look like:

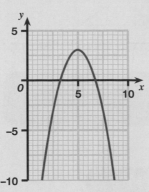

A quadratic function may also cross the x-axis twice (e.g. $y = x^2 - 6x + 5$), once (e.g. $y = x^2 - 4x + 4$) or not at all (e.g. $y = x^2 - 5x + 8$). However, it will always cross the y-axis at some point once.

Roots and **intercepts** of a quadratic function may be found graphically or algebraically.

This is the graph of
$y = x^2 - 2x - 3$

The graph crosses the x-axis twice, at $x = -1$ and $x = 3$. These are the roots of the equation $x^2 - 2x - 3 = 0$.
The intercept with the y-axis occurs at (0, –3).

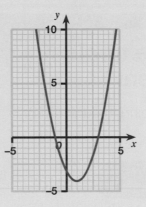

Consider the function $y = x^2 - 11x + 18$
To find the roots, factorise $x^2 - 11x + 18 = 0$
By factorising, you obtain $(x - 9)(x - 2) = 0$
$x - 9 = 0$ or $x - 2 = 0$
So the roots are $x = 9$ and $x = 2$

The curve crosses the y-axis when $x = 0$
So substituting $x = 0$ in the original equation: $y = 0^2 - 11 \times 0 + 18 = 18$.

Therefore, the graph will intercept the y-axis at (0, 18).

Turning points

Every quadratic equation has a **turning point** at a minimum (if $a > 0$) or maximum point (if $a < 0$).

Finding a turning point graphically

The turning point on the curve
$y = x^2 - 2x - 3$
is a minimum point and occurs at (1, –4).

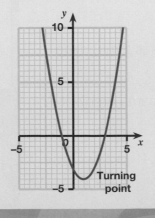

Turning point

The symmetrical property of a quadratic

Every quadratic graph $y = ax^2 + bx + c$ has a line of symmetry occurring at $x = -\dfrac{b}{2a}$

The roots also occur at an equal distance either side of the line $x = -\dfrac{b}{2a}$ ← It is also true that this is also the x-coordinate of the turning point on the curve.

Finding a line of symmetry

Find the equation of the line of symmetry of the curve $y = 3x^2 - 8x + 10$ ← $a = 3$ and $b = -8$

So the line of symmetry is $x = -\dfrac{b}{2a} = -\left(\dfrac{-8}{6}\right) = \dfrac{4}{3}$

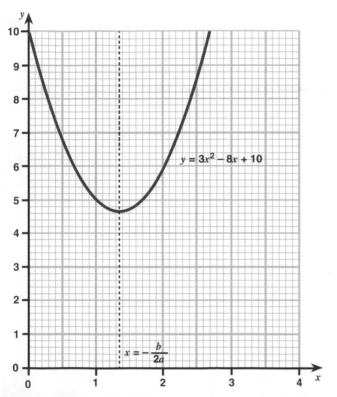

$y = 3x^2 - 8x + 10$

$x = -\dfrac{b}{2a}$

Draw a set of x-y axes on an A3 sheet of paper. Pick two specific points on the x-axis and one point on the y-axis. Draw a quadratic curve going through these points. Try to find the equation of the curve you have drawn.

1. Consider the curve $y = x^2 - 3x - 40$
   (a) Find the coordinates of the points where the curve intersects the x-axis.
   (b) State the coordinates where the curve crosses the y-axis.
   (c) Find the equation of the line of symmetry of the curve.
2. Consider the curve $y = x^2 + 10x - 40$
   (a) Find the equation of the line of symmetry of the curve.
   (b) Now find the coordinates of the turning point on the curve.

Functions

A **function** maps one number to another number.

Consider the function 'add 3'.

Input x		Output $f(x)$
1	maps to	4
2	maps to	5
3	maps to	6
4	maps to	7

It is represented graphically by plotting $y = x + 3$

This can be written as $f(x) = x + 3$

Consider the function 'square and subtract 2'.

Input x		Output $f(x)$
0	$\longrightarrow$	–2
1	$\longrightarrow$	–1
2	$\longrightarrow$	2
3	$\longrightarrow$	7

This can be written as $f(x) = x^2 - 2$
i.e. $f(0) = -2, f(1) = -1$, etc.

As well as **linear functions** $f(x) = ax + b$ and **quadratic functions** $f(x) = ax^2 + bx + c$, there are other types of function you need to know.

Cubic functions

Cubic functions have the form $y = ax^3 + bx^2 + cx + d$

The simplest cubic function is $y = x^3$

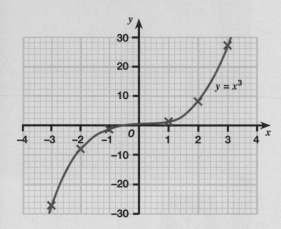

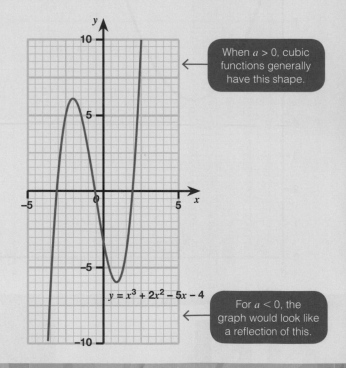

When $a > 0$, cubic functions generally have this shape.

$y = x^3 + 2x^2 - 5x - 4$

For $a < 0$, the graph would look like a reflection of this.

On A4 paper, draw an 'Input' column from 0 to 5, and a blank 'Output' column.

Try to find a function that has fewer than six different outputs for these inputs. Can you find more than one such function?

Module 16 · Other Functions

Reciprocal functions

Reciprocal functions have the form $y = \dfrac{k}{x}$ where k is a constant number. There are two types of curve, depending if k is positive or negative.

$y = \dfrac{2}{x}$

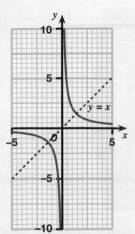

$y = \dfrac{-3}{x}$

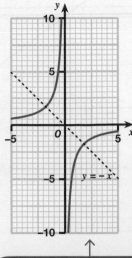

Each section of the graph approaches both axes and they are symmetrical about the line $y = x$

Each section of the graph again approaches both axes but they are in opposite quadrants to when $k > 0$. These graphs are symmetrical about $y = -x$

Function ➤ A type of mapping sending a number (input) to another number (output).

Linear function ➤ A mapping of the form $f(x) = ax + b$

Quadratic function ➤ A mapping of the form $f(x) = ax^2 + bx + c$

Cubic function ➤ A mapping of the form $f(x) = ax^3 + bx^2 + cx + d$

Reciprocal function ➤ A mapping of the form $f(x) = \dfrac{k}{x}$

1. For the function $f(x) = 5 - x$, write down the outputs for each of the following inputs:

Input x	Output $f(x)$
−1	→
0	→
1	→
2	→

2. Find the function given by the following mapping:

Input x	Output $f(x)$
1	→ 7
2	→ 9
3	→ 11
4	→ 13

3. Match the correct graph to each function:

$f(x) = \dfrac{2}{x}$ $\qquad$ $f(x) = \dfrac{1}{x}$ $\qquad$ $f(x) = \dfrac{-2}{x}$

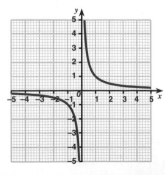

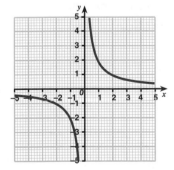

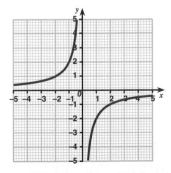

Practical problems involving graphs

This graph shows how the value of a new car changes over time:

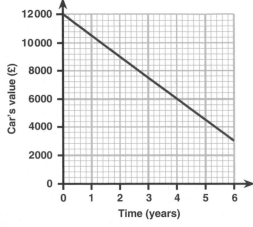

Find the **rate of depreciation** of the car.

In six years, the car's value reduces from £12 000 to £3000, a difference of £9000.

The rate of depreciation is therefore $\frac{9000}{6}$ = £1500 per year.

The opposite of a rate of depreciation is a **rate of appreciation**.

Distance–time and speed–time graphs

In a **distance–time graph**, the gradient represents the **speed** at a given time.

The graph opposite shows Max travelling to his friend's house.

(a) Find the speed at which Max walks to his friend's house.

The gradient of the first line is $\frac{1.5}{0.5}$ = 3km/h

(b) Find the speed at which he returns.

The gradient of the final line is $\frac{1.5}{2}$ = 0.75km/h

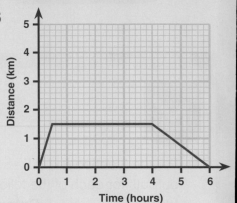

In a **speed–time graph**, the gradient represents the **acceleration** at a given time.

The area under a speed–time graph represents the distance the object has travelled.

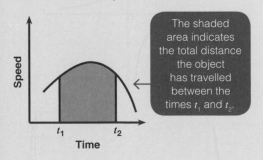

The shaded area indicates the total distance the object has travelled between the times t_1 and t_2.

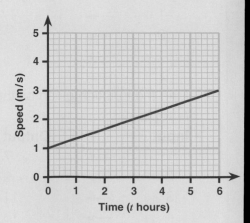

The graph above right shows the speed of an object at time t.

(a) Find the acceleration of the object.

The gradient of the line is $\frac{2}{6} = \frac{1}{3}$

So the acceleration is $\frac{1}{3}$m/s²

(b) Find the total distance travelled by the object.

The area under the graph is that of a trapezium.

The area is $\frac{1}{2} \times (1 + 3) \times 6 = 12$

So the total distance travelled is 12m.

Module 17 Problems and Graphs

An object travels in a straight line from rest such that its speed s at time t is given by $s = t \times 2^{-t}$ for $t \geqslant 0$.

The graph of s against t is shown.

(a) Find the speed of the object after 4 seconds.

From the graph, you can see that when $t = 4$, $s = 1\text{m/s}$.

(b) Find the time(s) when the speed of the object is 2m/s.

By drawing a straight line across from $s = 2$, you can see that this intersects the curve in two places, namely when $t = 0.6\text{s}$ and $t = 2\text{s}$

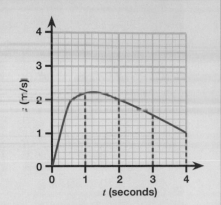

1. The graphs A, B, C, D and E show how the populations have changed in five different countries. Match each country to its description.

   1 – The population increased steadily over time.
   2 – The population remained steady and did not increase or decrease over time.
   3 – The population decreased at a quick rate after decreasing quite slowly.
   4 – The population increased slowly at first but then increased at a quicker rate.
   5 – The population decreased at a constant rate.

2. The graph opposite shows the speed of a cyclist over 15 seconds.

   (a) For how long is the cyclist travelling at a constant speed?
   (b) Calculate the deceleration of the cyclist.
   (c) Calculate the total distance travelled by the cyclist.

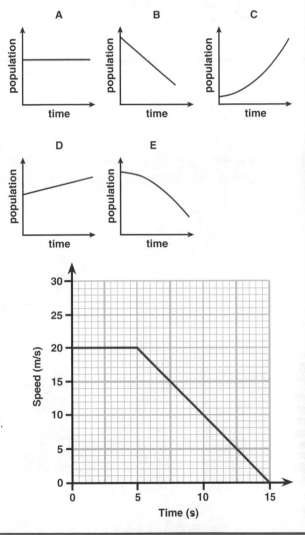

Imagine a skydiver jumping out of an aeroplane. After falling halfway to the ground, he opens his parachute and continues to fall, before safely reaching the ground. On a large sheet of paper, draw a set of axes, marking the x-axis as time and the y-axis as speed. Try to draw the speed–time graph for the skydiver, marking the point where the skydiver opens his parachute.

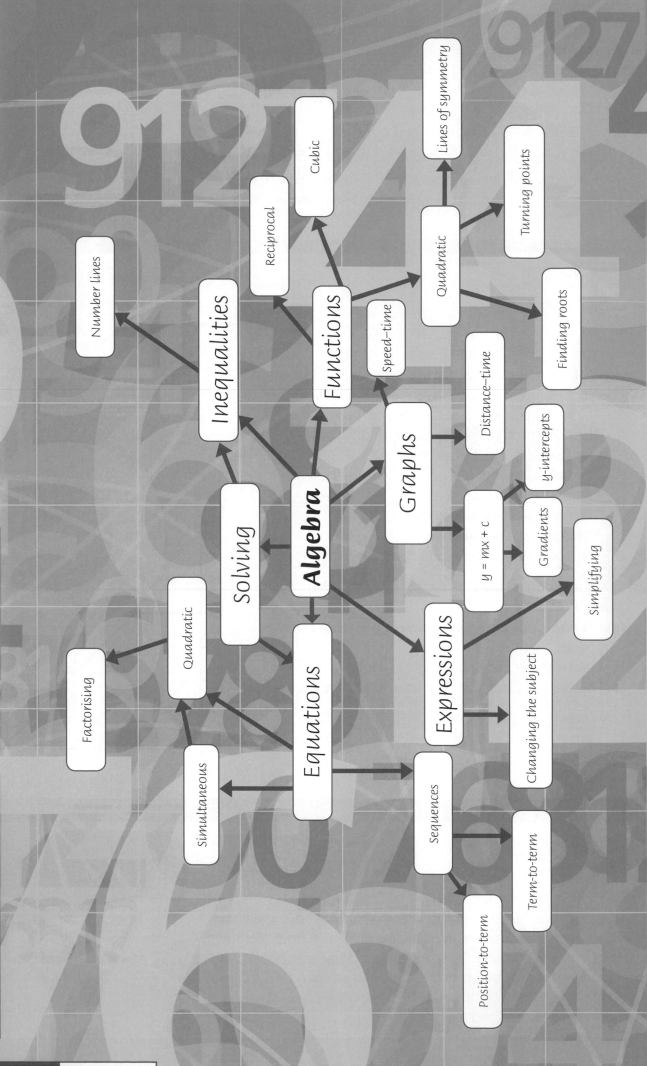

1. Solve the following equations.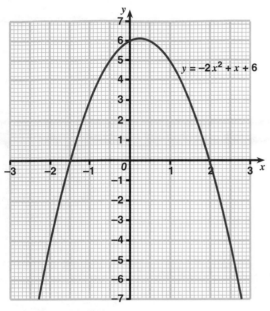

   (a) $3(x - 2) - 4(x - 1) = 0$ [2]

   (b) $\dfrac{x + 8}{3} = \dfrac{2x - 9}{5}$ [3]

2. Which of the following lines are parallel? [2]

   A: $\boxed{3y = x - 6}$ **B:** $\boxed{4x + 2y = 8}$

   C: $\boxed{x = 3y - 12}$ **D:** $\boxed{2y - x + 2 = 0}$

3. Given the formula $S = 6r^2h$, work out the value of S when $r = 2$ and $h = 0.5$ [2]

4. Factorise fully $18a^2 - 6ab$ [2]

5. The following graph shows a quadratic curve $y = f(x)$

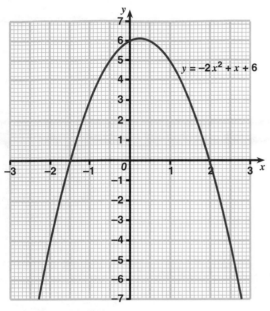

$y = -2x^2 + x + 6$

   From the information in the graph, write down:

   (a) The roots of $f(x) = 0$ [2]

   (b) The coordinates of the y-intercept. [1]

   (c) The equation of the line of symmetry. [1]

6. Solve the following equations by factorisation:

   (a) $x^2 + 16x + 63 = 0$ [3]

   (b) $x^2 + 7x - 30 = 0$ [3]

7. Find the next term in the following sequence.

   5   21   47   83   129 [2]

8. If $y = \dfrac{3x + 1}{x - 2}$, make x the subject. [3]

9. Solve the simultaneous equations $2x + 3y = 1$
   $$3x - 2y = -18$$ [4]

10. **(a)** Sketch the curve of $y = \dfrac{4}{x}$ [2]

    (b) What straight line should you draw on the same axes in order to
    solve the equation $x^2 + x = 4$? [3]

Metric and imperial units

In the UK we still use a mix of different units and you need to know some standard conversions between the two systems.

Length	2.5cm ≈ 1 inch	8km ≈ 5 miles
Volume	1 litre ≈ $1\frac{3}{4}$pts	4.5 litres ≈ 1 gallon
Mass	1kg ≈ 2.2lb	28g ≈ 1 ounce

Time

60s = 1 minute
60 mins = 1 hour
24h = 1 day
365 days = 1 year

Be very careful with time. Your calculator may give an answer of 2.25 hours. This is **not** 2h 25min but $2\frac{1}{4}$ h which is 2h 15min. If you have to enter 3h 30min into your calculator, this is 3.5h **not** 3.30h.

The metric system

Metric measures are based on units of 10.

milli... means $\frac{1}{1000}$	*centi*... means $\frac{1}{100}$	*kilo*... means 1000

Mass: 1000g = 1kg   and   1000kg = 1 tonne
Length: 1000mm = 100cm = 1m   and   1000m = 1km
Capacity: 1000ml = 1 litre   and   1000 litres = 1m³

Currency conversions

You must always use your common sense. Conversions are always given as a ratio, e.g. £1 : $1.60

Imagine the bank taking away your £1 coin and giving you $1.60 instead. You have more dollars than you had pounds (although they are worth the same).

Convert £450 to $ 450 × 1.6 = $720   ← More dollars than pounds

Convert $620 to £   620 ÷ 1.6 = £387.50   ← More dollars than pounds

Compound units

You need to know about speed, density, pressure and rates of pay.

Speed = $\frac{distance}{time}$ so the units will always be 'distance' per 'time'.

metres per second	centimetres per second	km per hour	miles per hour	feet per second
m/s or ms⁻¹	cm/s or cms⁻¹	km/h or kmh⁻¹	mph	ft/s or fts⁻¹

Density = $\frac{mass}{volume}$ so the units will always be 'mass' per 'volume', e.g. kg/m³ (or kgm⁻³) and g/cm³ (or gcm⁻³).

Pressure = $\frac{force}{area}$ (units include Newtons/m²)

Rates of pay are given for a unit of time so £18/h or £500/week or £4400/month or £33 000/annum.

Converting between **compound units** should always be done in steps.

Change 80km/h into m/s.

	80km	in 1 hour
is	80 000m	in 1 hour
is	80 000m	in 60min
is	1333.333...m	in 1min
is	22.22...m	in 1s

So 80km/h is 22.2m/s (3 s.f.)

Conversions for area and volume

A diagram always helps when converting between different units for area and volume.

Area

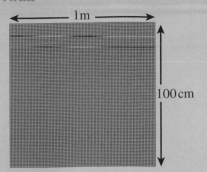
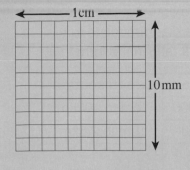
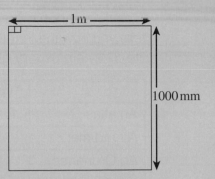

Filling a 1m² with cm²
➤ 100 rows of 100
➤ 10 000cm² (10⁴ cm²) in 1m²

Filling a 1cm² with mm²
➤ 10 rows of 10
➤ 100mm² in 1cm²

Filling a 1m² with mm²
➤ 1000 rows of 1000
➤ 1 000 000mm² (10⁶ mm²) in 1m²

Volume

Imagine filling a box 1m × 1m × 1m with tiny cubes each 1cm × 1cm × 1cm. There would be 100 rows of 100 on the bottom layer and 100 layers. There are 1 000 000cm³ in 1m³.

Imagine filling a box 1cm × 1cm × 1cm with tiny cubes each 1mm × 1mm × 1mm. There would be 10 rows of 10 on the bottom layer and 10 layers. There are 1000mm³ in 1cm³.

Imagine filling a box 1m³ with mm³. There would be 1000 rows of 1000 and 1000 layers. There are 1 000 000 000mm³ (10⁹) in 1m³.

Measure some water into a measuring jug. Weigh the jug of water and then weigh the jug empty. Work out the weight of the water and calculate its density in either g/cm³ or kg/m³.

18

1. Convert 64km/h to m/s.
2. Amil sees some jeans for sale in France for €80. If they cost £48 in the UK, where should he buy them? (£1 = €1.25)
3. Convert 8400mm² to cm².
4. Convert 2.5m³ to cm³.

Scale factors

You will meet scale factors in **scale drawings**, **similar triangles** and in **enlargements**.

A scale diagram is an accurate drawing, the scale tells you how the lengths in the diagram relate to real life. They are usually given as a ratio in the form 1: n

19

	Typical scale	…represents…
An atlas	1:10 000 000	1cm to 100km
A road map	1:250 000	1cm to 2.5km
An Ordnance Survey Explorer map	1:25 000	1cm to 250m
An architect's plan	1:100	1cm to 1m
A modeller's plan	1:24	1cm to 24cm

Scale factors and diagrams

If you have to calculate a distance from a map or plan:

➤ Measure the distance accurately with a ruler
➤ Write down the scale given
➤ Calculate the scaled-up 'real' size.

If you have to draw a map or plan:

➤ Decide on a suitable scale
➤ Write down the 'real' distances
➤ Calculate the scaled-down lengths
➤ Draw the plan accurately.

Beware!

The angles are always the same in both the plan and real life.

The areas on a scale drawing will be reduced by [scale factor]^2 – see Module 18.

Scale 1 : 20 000

The distance between the Round Pond and Lancaster Gate station is 4cm.

In real life this is 800m.

Module 19 Scales, Diagrams and Maps

Bearings

Use bearings to set an absolute direction, measured from compass North. You always start at a point with your toes facing North. Then you turn clockwise until your toes are pointing in the direction that you want to go.

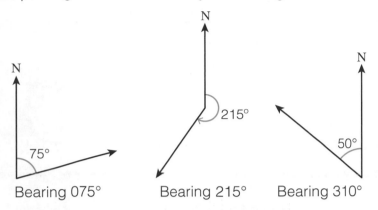

Bearing 075°   Bearing 215°   Bearing 310°

Use a compass to find North. Your school might have a 'bearings compass' or you might find one on a phone or digital watch. Mark this direction with a ruler or something similar and then try to find the bearing of some objects, for example the garden shed or a lamp-post.

If the bearing is less than 100° you insert a leading zero – all bearings must have three figures.

When you turn around and walk back the way you came, you turn through 180°.
If you have to reverse the direction of a bearing (a 'back bearing'), you simply add 180° to its value.

Bearing	15°	95°	110°	285°
Back bearing	15 + 180 = 195°	95 + 180 = 275°	110 + 180 = 290°	285 + 180 = 465° More than one complete turn so 465 − 360 = 105°

Bearings questions:
- ➤ can ask you to measure accurately or draw angles   (Module 24)
- ➤ can involve angles and parallel lines   (Module 25)
- ➤ can include Pythagoras' theorem and trigonometry.   (Module 32)

Ahab's sat-nav is broken but he has a compass and is sailing North. It is dark and he knows there are dangerous rocks. Coastguards A and B standing on the cliff can see his lights and measure his bearings as 160° and 230°.

Draw bearings and decide if he will hit the rocks if he carries on going North.

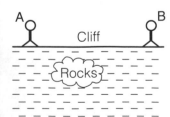

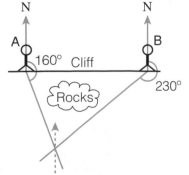

Ahab's position is where the two bearing lines cross. If he keeps going North, he will hit the rocks.

1. The scale of a road atlas is 1 : 250 000
   What do these distances represent in real life (in km)?
   (a) 4cm   (b) 10cm   (c) 1.5cm
   How far would these distances be on the map (in cm)?
   (d) 35km   (e) 40km   (f) 5km

2. Draw diagrams to show bearings of
   (a) 085°   (b) 190°   (c) 300°
   What is the back bearing in each case?

Fractions, decimals and percentages

Per cent means 'out of 100'. There is a direct link between fractions, decimals and percentages.

Fraction	$\dfrac{n}{100}$	$\dfrac{1}{100}$	$\dfrac{50}{100}=\dfrac{1}{2}$	$\dfrac{25}{100}=\dfrac{1}{4}$	$\dfrac{75}{100}=\dfrac{3}{4}$	$\dfrac{10}{100}=\dfrac{1}{10}$	$\dfrac{5}{100}=\dfrac{1}{20}$	$\dfrac{120}{100}=1\tfrac{1}{5}$
Decimal		0.01	0.5	0.25	0.75	0.1	0.05	1.2
Percentage	$n\%$	1%	50%	25%	75%	10%	5%	120%

Calculating with percentages

There are many correct ways to calculate with percentages. The best method often depends whether you are allowed to use a calculator or not.

Three ways to calculate 'increase 480 by 21%' are shown.

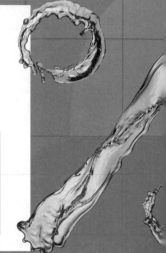

Without a calculator	With a calculator	With a calculator and using a multiplier
1% of 480 $= 4.8$ 10% of 480 $= 48$ 20% of 480 $= 96$ $480 + 96 + 4.8 = 580.8$	21% of 480 $=$ $\dfrac{21}{100} \times 480 = 100.8$ $480 + 100.8 = 580.8$	New amount is 121% of original. ← 'of' means '×' $1.21 \times 480 = 580.8$

Comparisons using percentages

Percentages allow comparisons to be made easily as everything is written as parts of 100.

If you got 15 in a maths test, you can't tell how well you did unless you know the maximum possible marks. In each case write 15 as a fraction of the total available marks and then write it as a percentage.

$$15 \text{ out of } 20 = \frac{15}{20} = 0.75 = 75\%$$

$$15 \text{ out of } 30 = \frac{15}{30} = 0.5 = 50\%$$

$$15 \text{ out of } 50 = \frac{15}{50} = \frac{30}{100} = 30\%$$

Equivalent fractions

KEYWORDS

Depreciate ➤ When an object loses value (e.g. cars, motorbikes, computers).

Appreciate ➤ When an object gains value (e.g. property, investments, gold).

Per annum ➤ Per year.

Compound (interest) ➤ The interest earned added so the sum invested increases each year.

Percentages and money

An article can **appreciate** (increase) or **depreciate** (decrease) in value by a given percentage.

➤ The value of a painting has appreciated by 15%. If its value **was** £45 000, it is now worth $1.15 \times 45\,000 = £51\,750$

➤ The value of a car has depreciated by 15%. If its value **was** £45 000, it is now worth $0.85 \times 45\,000 = £38\,250$

  If it lost 15%, it would be now worth 85%

➤ If I bought a mansion for £2.5 million and sold it for £3.2 million:
  My **actual** profit would be £0.7 million or £700 000.
  My **percentage** profit would be
  $$\frac{0.7}{2.5} = \frac{7}{25} = \frac{28}{100} = 28\%$$

Simple and compound interest

When you invest a sum and then take the interest to spend (perhaps using the interest to live on), this is simple interest. Every year begins with the same sum invested to earn interest.

When you invest a sum and leave the interest in the account (to **compound**), each year the sum is larger and so the interest earned increases. It is easiest to calculate using a multiplier.

Simple	Compound	
£500 at 2% **per annum**	£500 at 2% per annum	Using a multiplier
Y1 500 × 0.02 = 10	Y1 500 × 0.02 = 10	Y3 = 500 × 1.02^3 = 530.60
Y2 500 × 0.02 = 10	Y2 (500 + 10) × 0.02 = 10.20	Total interest
Y3 500 × 0.02 = 10	Y3 (510 + 10.20) × 0.02 = 10.40	530.60 − 500 = £30.60
Total interest £30	Total interest	
	10 + 10.20 + 10.40 = £30.60	

The compound interest principle also applies to repeated depreciation.

If a car worth £10 000 depreciates by 15% per year for five years, then its value is:

10 000 × 0.85^5 = £4437

↑
It loses 15% so 85% is left.

Make a drink of fruit cordial and measure how much concentrate and how much water you use. Then calculate the percentage of cordial in the drink.

Check the label on the bottle. Can you work out what percentage of your glass contains sugar?

Working backwards

The more difficult questions ask you to work backwards to find a value **before** an increase or decrease. You have to be very careful to read the question correctly, choose your method and show all your steps clearly.

  20

A laptop costs £348.50 in a '15% off' sale. What was its price before the sale?

Using %	Using a multiplier
The sale price is 85% of the original. $\frac{348.50}{85}$ = 4.1 (1% of the original). 4.1 × 100 = £410	The sale price is 85% of the original. 85% = 0.85 $\frac{348.50}{0.85}$ = 410 so £410

A price increase of 8% means a new car is now £27 000. What was its price before the increase?

Using %	Using a multiplier
The new price is 108% of the original. $\frac{27 000}{108}$ = 250 (1% of the original). 250 × 100 = £25 000	The new price is 108% of the original. 108% = 1.08 $\frac{27 000}{1.08}$ = 25 000 so £25 000

1. Without a calculator, work out these.
   (a) 15% of £56
   (b) Reduce 560 by 30%
   (c) The value of a £16 000 car after value added tax (VAT) of 20% is added.
2. Using a calculator, work out these.
   (a) 17% of £112
   (b) Increase 456 by 27%
   (c) 42 as a percentage of 520
3. If a plumber's bill was £460.80 after VAT at 20%, what was the price without VAT?

Writing ratios

You can use ratio, using colons (:), to show the proportion between two or more numbers.

If there are 25 dogs and 45 chickens, the ratio of dogs to chickens is
    25 : 45 or
     5 : 9 in its simplest form. ← Divide both parts by 5.
But the ratio of dogs' feet to chickens' feet is
    100 : 90 or 10 : 9 in its simplest form.

If there are 25 dogs, 45 chickens and 20 sheep, the ratio of dogs : chickens : sheep is
    25 : 45 : 20 or
     5 : 9 : 4 in its simplest form.

Units should be the same when you make comparisons.
£2.50 : 50p as a ratio is 250p : 50p = 5 : 1
5m : 7000cm = 5m : 70m = 1 : 14

Unitary form (1 : n) is most commonly used in maps and scale drawings.
2 : 3 in unitary form is 1 : 1.5
4 : 5 in unitary form is 1 : 1.25
5 : 4 in unitary form is 1 : 0.8

Ratio and fractions

Reducing a ratio to its simplest form is similar to cancelling fractions.

16 : 18 simplifies to 8 : 9 ← Divide both parts by 2.

21 : 28 simplifies to 3 : 4 ← Divide both parts by 7.

There is a link between using ratio and fractions to divide amounts:

Ratio	1 : 2	2 : 3	3 : 5 : 8	1 : 1	1 : 4
Parts / Lots	3	5	16	2	5
Fractions	$\frac{1}{3}, \frac{2}{3}$	$\frac{2}{5}, \frac{3}{5}$	$\frac{3}{16}, \frac{5}{16}, \frac{8}{16}\left(=\frac{1}{2}\right)$	$\frac{1}{2}, \frac{1}{2}$	$\frac{1}{5}, \frac{4}{5}$

 Cut out 20 pictures of people, including both children and adults, from a newspaper or magazine. Organise your pictures into two groups – males and females – and write down the ratio of males to females in its simplest form. Then organise the pictures into children and adults and write down the ratio of children to adults in its simplest form.

Using ratio

To divide an amount in a given ratio, you first need to work out how many parts there are in the ratio.

Divide £840 in the ratio 6 : 5 : 1

There are 6 + 5 + 1 = 12 lots

$\frac{840}{12} = 70$ so 1 lot is £70

6 × 70 = 420; 5 × 70 = 350; 1 × 70 = 70

So £420 : £350 : £70 ← Check your answer by adding 420 + 350 + 70 = 840

Always read the question carefully.

Jane and Peter share a basket of apples in the ratio 3 : 5
If Peter had 15 apples, how many were there altogether?

Peter's apples are 5 parts of the ratio so

1 part is $\frac{15}{5}$ = 3 apples.

The whole basket was 8 parts, which is 3 × 8 = 24 apples.

There are different ways to compare the value of items.

Which is the best value?

500g £3.50

700g £4.75

For maximum marks, set out your working clearly.

Method 1 Find a price for equal amounts.

Jar A	Jar B
500g cost £3.50	700g cost £4.75
100g cost £0.70	100g cost £0.68

Jar B cheaper for 100g.

Method 2 Find how much you get for an equal price.

Jar A	Jar B
£3.50 buys 500g	£4.75 buys 700g
£1 buys 142.9g	£1 buys 147.4g

£1 buys more with Jar B so Jar B cheaper.

Method 3 Use common multiples.

Jar A	Jar B
500g cost £3.50	700g cost £4.75
× 7	× 5
3500g cost £24.50	3500g cost £23.75

Jar B cheaper for equal amounts.

1. Peter, Quinn and Robert have been left £44 800. They share it in the ratio of 3 : 5 : 6.
   How much does Quinn get? What fraction of the money does Robert get?
2. Which is the best value for money, 7.5m of ribbon for £6.45 or 5.4m for £4.43?
3. Farmer Jones has 80 chickens. If the ratio of chicken's feet : sheep's feet is 2 : 3, how many sheep does he have?

Direct proportion

When two things are in a constant ratio, they are **directly proportional** (∝) to one another – as one increases so does the other.

There are many examples of direct proportion but the best is probably buying goods. The more bags of cement you buy, the more you will pay.

The total cost (C) is directly proportional to the number of items (n).

Inverse proportion

When two things are **inversely proportional** to one another, one goes up as the other goes down.

Sharing sweets is a good example of inverse proportion.

The more people that share, the fewer sweets they each get.

24 sweets to share: Two people get 12 each or three people get 8 each or twelve people get 2 each.

The number of sweets (s) is inversely proportional to the number of people (p).

22

Expressing proportions

Words	Symbols	Equation (k is a constant)
A is directly proportional to B	$A \propto B$	$A = kB$
A is inversely proportional to B	$A \propto \dfrac{1}{B}$	$A = \dfrac{k}{B}$

Solving proportion problems

Variables change in proportion in many areas of maths (and science) including problem solving, similarity, transformations and trigonometry.

Problem-solving

Temperature (T) is inversely proportional to pressure (P). If T is 5.5 when P is 8, what is T when P is 4?

$T = \dfrac{k}{P}$ so $5.5 = \dfrac{k}{8}$

$\Rightarrow k = 5.5 \times 8 = 44$

$\Rightarrow T = \dfrac{44}{P}$

When $P = 4$,

$T = \dfrac{44}{4} = 11$

KEYWORDS

$\propto$ ➤ The symbol for 'proportional to'.

Directly proportional ➤ Variables that are in a fixed ratio.

Inversely proportional ➤ As one variable increases, the other decreases.

Similar shapes ← Also see page 62.

A sheet of A4 paper, when measured to the nearest mm, is 210mm wide by 297mm long. A sheet of A5 is mathematically similar. If it is 140mm wide, what is its length?

Length $\propto$ width so $L = kW$

$297 = k\,210 \Rightarrow k = \dfrac{297}{210} = \dfrac{99}{70}$

$\Rightarrow L = \dfrac{99}{70}W$

When $W = 148$mm $L = \dfrac{99}{70} \times 148 = 209$ so 209mm

This problem could also have been solved using ratios.
If the shapes are similar then the ratio between corresponding sides is the same (one is an enlargement of the other).

$210 : 297$

$= 1 : \dfrac{297}{210}$ ← Divide by 210 (leave as a fraction).

$= 148 : 209$ ← Multiply by 148.

Cut a sheet of A4 paper in half to make two pieces of A5 size. Cut one A5 piece in half again to make A6. How many A6 pieces would fit on an A4 sheet? If you can find A3 paper then cut that to make A4, A5 and A6. How many A6 sheets would fill one A3 sheet?

1. If 4m of wire costs £6.60, how much will 7m cost?
2. P is proportional to Q and P is 7 when Q is 12. Find the value of P when Q is 6.
3. C is inversely proportional to D. If $C = 9$ when $D = 5$, find D when $C = 3$.

Rates of change and graphs

This graph shows the cost of a mobile phone on three different tariffs.

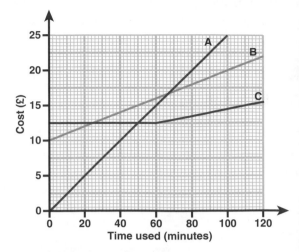

➤ Line A shows direct proportion. The **gradient** of the line is the rate of change. $\dfrac{\text{change in } y}{\text{change in } x} = \dfrac{10}{40} = 0.25$
This is the constant of proportion and represents the cost per minute of 25p.

➤ Line B shows a fixed charge of £10 before any calls are made. The gradient is $\dfrac{5}{50} = 0.1$
This represents a cost per minute of 10p.

➤ Line C shows a fixed charge of £12.50 and all calls free up to a total of 60 minutes.
After that the gradient is $\dfrac{2.5}{50} = 0.05$
This represents a cost per minute of 5p for all calls after 60 minutes have been used.

You can use this graph to choose which is the best tariff for you. For more than 50 minutes use, C is cheaper, but for low use A is best.

This graph shows Paul's journey from home on foot and by car.

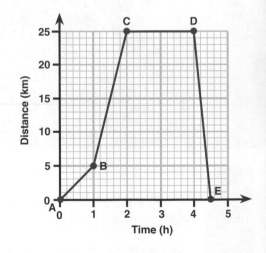

➤ From A to B Paul is walking – he travels 5km in 1 hour.

➤ From B to C Paul is in a car – he travels 20km in 1 hour so his speed is 20km/h.

➤ From C to D Paul is in the same place, perhaps visiting friends.

➤ From D to E Paul returns home. He travels 25km in half an hour, so his speed is 50km/h. He finishes his journey back at home.

Take 25 counters (or make your own using pieces of paper). Increase the number of counters by 20%. Increase this number by 40%. Reduce your latest number of counters by 50% and write down how many you are left with.

Inverse proportion graphs

A graph showing inverse proportion looks like this – it is hyperbolic (a **hyperbola**).

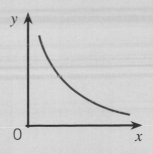

Repetitive rates of change

When a percentage change is applied more than once, you must take care to do each step at a time.

A new car costs £23000. In the first year it depreciates by 20%, the next year it loses 15% and then 10% every year after that. To work out the value of the car after three years:

Step-wise
Lose 20% so worth 80%
80% of 23000 = 0.8 × 23000 = 18400
then
85% of 18400 = 0.85 × 18400 = 15640
then
90% of 15640 = 0.9 × 15640 = 14076
Worth £14076 after three years
Using multipliers
Lose 20% so worth 80%
Lose 15% so worth 85%
Lose 10% so worth 90%
0.8 × 0.85 × 0.9 = 0.612
0.612 × £23000 = £14076

1. Work out the gradient of these straight lines.

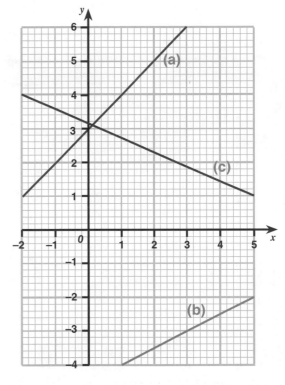

2. These graphs show the height of liquid in a jar as water is poured in at a constant rate.

Match each graph to the correct container.

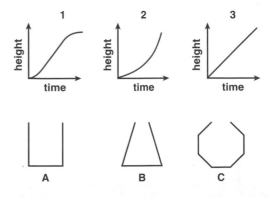

Ratio, Proportion and Rates of Change Mind Map

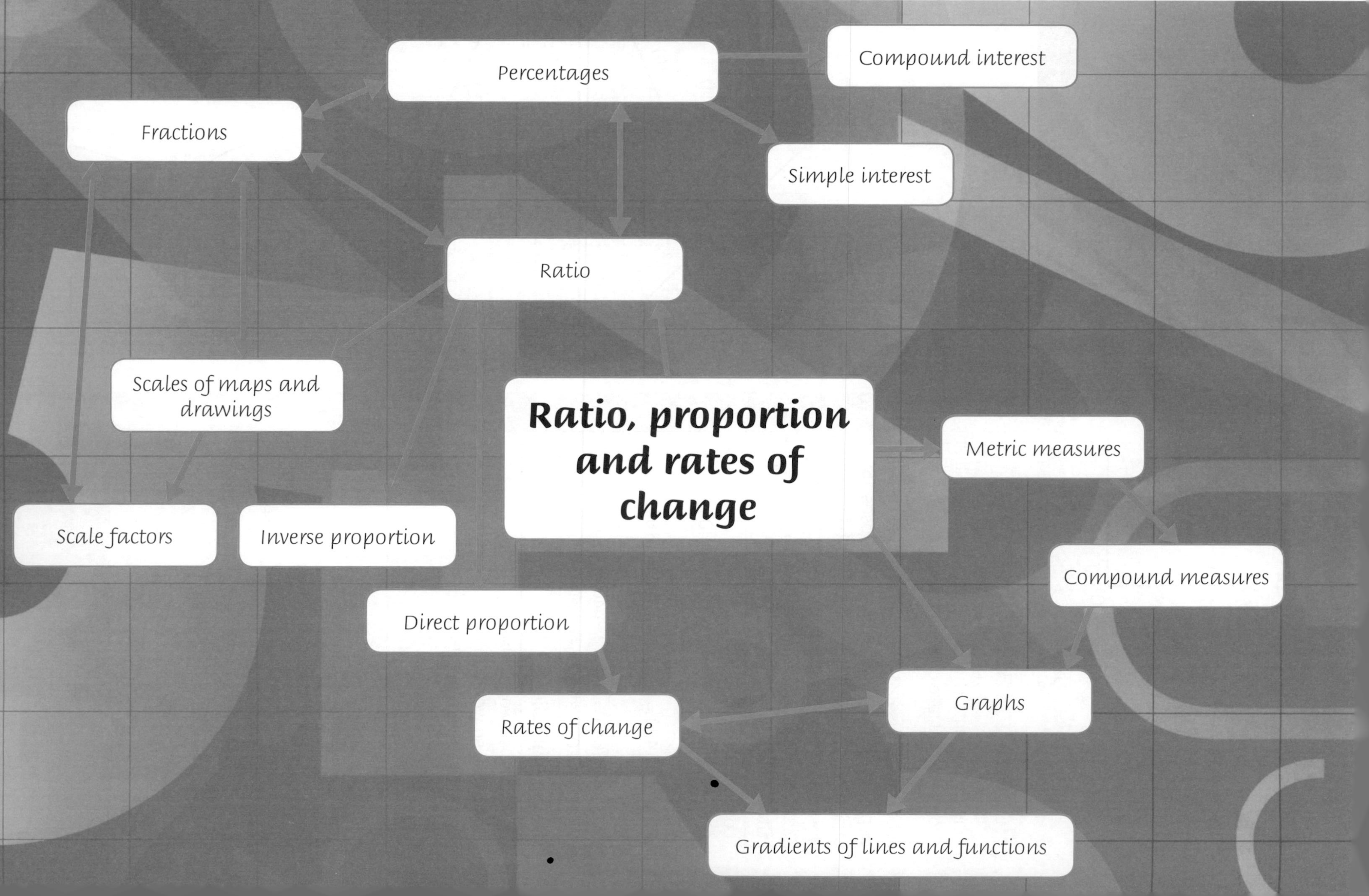

Ratio, proportion and rates of change
Percentages
Compound interest
Simple interest
Fractions
Ratio
Scales of maps and drawings
Scale factors
Inverse proportion
Direct proportion
Rates of change
Metric measures
Compound measures
Graphs
Gradients of lines and functions

1. Convert 2.5m² to cm².  [2]

2. Alpha, Beta and Gamma share a legacy in the ratio 3 : 4 : 5

   If Beta had £2400, how much was left to them altogether?

   What fraction of the whole sum did Gamma inherit?  [4]

3. Amil is buying a computer game on the Internet.

   If the exchange rate is £1 to €1.22, which should he buy? How much
   will he save?

   [3]

4. James and Julia bought a house for £824 000 in 2011. It lost value and they sold
   it for £760 000 in 2015.

   What was their percentage loss on the transaction? [2]

5. A map is drawn using a scale of 1 : 200 000
   Peter measures the motorway distance between junctions 9 and 11 as 9cm.

   (a) What is the actual distance between junctions 9 and 11 in kilometres? [2]

   (b) If he drives at an average speed of 90km/h, how many minutes will this
   part of his journey take him? [2]

6. The graph shows the position of a particle over a period of 14 seconds.

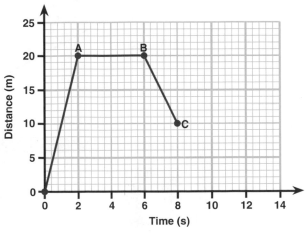

   (a) Find the speed of the particle between 0 and A. [2]

   (b) What happens between A and B? [1]

   (c) After eight seconds the particle returns to its original position at a speed
   of 2.5m/s. Complete the graph. [2]

Labelling and drawing

You must be able to draw lines of a given length accurately, and use a protractor to draw and measure angles to the nearest degree.

You will see triangles and angles labelled in many different ways.

This triangle could be called triangle *ABC*, *BCA* or *CBA*.

The shaded angle could be correctly called the **acute angle** at *A* or just '*a*':

angle *BAC*   or   angle *CAB*

∠*BAC*   or   ∠*CAB*

BÂC   or   *CÂB*

The side *CA* is extended to *D*.

The **obtuse angle** *DÂB* is an external angle. **Reflex angles** are between 180° and 360°.

Locus

Locus is a Latin word (plural loci) which means location or place. In geometry the locus of a set of points is all the places it can be, according to a given rule.

24

The locus of all points 3cm from the point *C*.	The locus of all points **equidistant** from the **line segment** *AB*.	The locus of all points equidistant from the points *P* and *Q*.	The locus of all points equidistant from the lines *FG* and *GH*.
A circle radius 3cm, centre *C* (or in 3D a sphere).	Two parallel lines joined with semicircles.	The **perpendicular bisector** of the line *PQ*.	The bisector of ∠*FGH*.
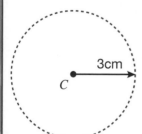	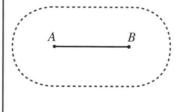	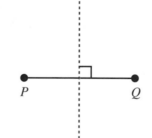	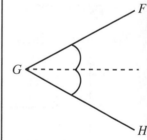

Acute angle ➤ An angle between 0° and 90°.

Obtuse angle ➤ An angle between 90° and 180°.

Equidistant ➤ The same distance from.

Line segment ➤ A short section of an infinite line.

Perpendicular to ➤ At 90° to.

Bisector ➤ Divides in half.

Locus/loci ➤ A set or sets of points.

KEYWORDS

Chalk a line segment on the ground. Using a tape measure or string to help you, chalk the locus of all the points 1 metre from the line segment.

Module 24 **Constructions**

Accurate constructions

You may have to construct angles and shapes without using a protractor to measure angles. You will only be allowed to use a ruler and a pair of compasses. ←

You need to know that:
➤ a rhombus has diagonals that bisect its angles
➤ an isosceles triangle has a line of symmetry that bisects the base
➤ an equilateral triangle has all angles = 60°
➤ the shortest distance drawn from a point to a straight line is perpendicular to the line.

1. **The locus of all points equidistant from P and Q**

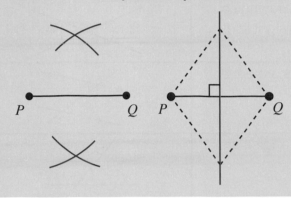

➤ You will construct two isosceles triangles with PQ as the base.
➤ Set compasses to a distance about the same as PQ.
➤ Use arcs to mark the vertices above and below PQ.
➤ Join the vertices – the line is the perpendicular bisector of PQ.

2. **The locus of all points equidistant from the lines FG and GH**

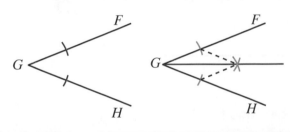

➤ You will draw a line that bisects the angle by constructing a rhombus.
➤ Set compasses to a fixed distance.
➤ From G mark two equal lengths on the lines FG and GH.
➤ Using these marks set the fourth vertex of the rhombus.
➤ Draw the diagonal. This is the angle bisector.

3. **Angles of 60° and 30°**
➤ Using a pair of compasses and a ruler, construct **any** equilateral triangle. This gives an angle of 60° **without** using a protractor.
➤ To draw 30° you need to bisect the 60° angle. Construct a rhombus – see **2**.

4. **Angles of 90° and 45°**
➤ Using a pair of compasses, construct a line and its perpendicular bisector to make 90° – see **1**.
➤ To draw 45° you need to bisect the 90° angle. Construct a rhombus – see **2**.

1. Draw the following:
   (a) The locus of all points 3cm from a line of length 5cm.
   (b) The locus of all points equidistant from points S and T which are 6cm apart.
   (c) Two intersecting lines AB and BC and the locus of all points equidistant from both lines.
2. Construct an angle of 30° using a pair of compasses and a ruler only.

Angle rules

Look at these angle rules and examples:

Angles at a point	The angles around a point add up to 360°.	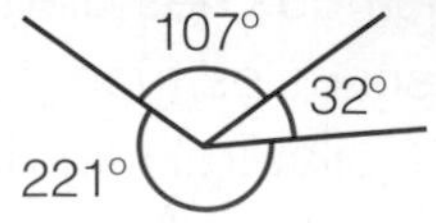 107° 32° 221°
Angles on a straight line	The angles along a straight line add up to 180°.	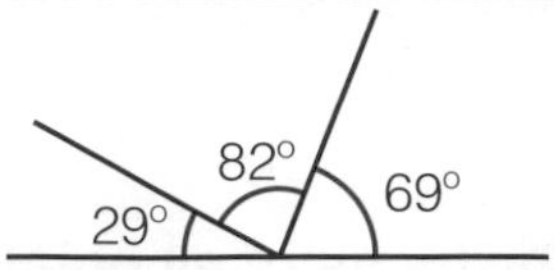 82° 29° 69°
Vertically opposite angles	Vertically opposite angles between two straight lines are equal.	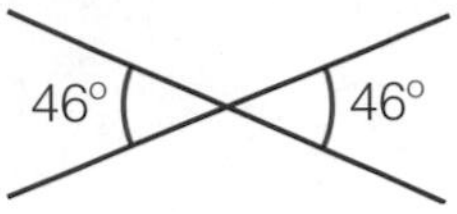 46° 46°
Parallel lines and alternate angles	Alternate angles are equal.	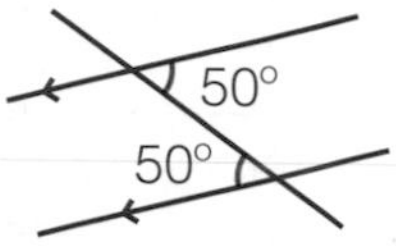 50° 50°
Parallel lines and corresponding angles	Corresponding angles are equal.	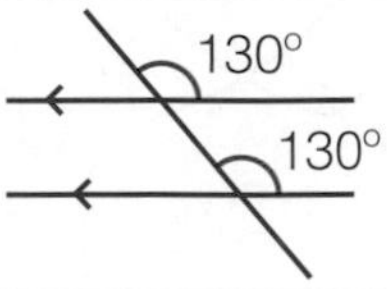 130° 130°
Angles in a triangle	Angles in a triangle always add up to 180°.	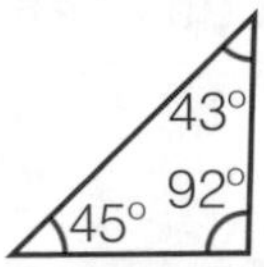 43° 92° 45°

The properties of triangles

The shape of a triangle is described using its angles and lengths.

Name	Sides	Angles	
Equilateral	Three sides equal; a **regular** triangle	Three angles equal (60°)	Has three lines of symmetry 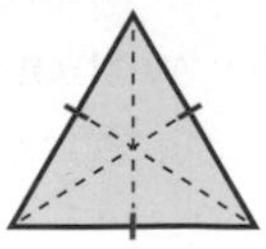
Isosceles	Two sides equal	Two angles equal	Has one line of symmetry 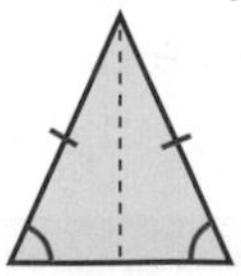
Scalene	No sides equal	No angles equal	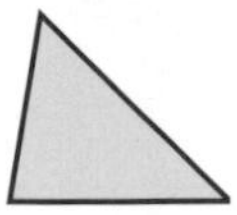
Right-angled	Could be isosceles or scalene	One right angle	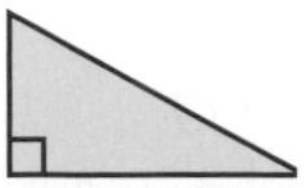 
Obtuse-angled	Could be isosceles or scalene	One obtuse angle	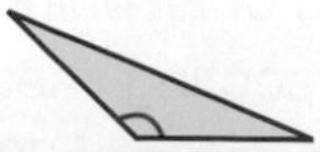 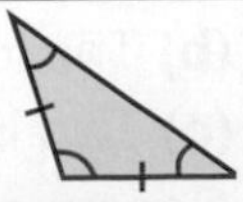

Angles

Module 25

Proof of angle sum for any triangle

You can use the angle rules to **prove** that the angles inside a triangle always add up to 180°.

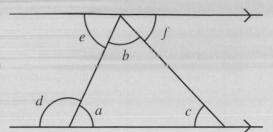

$e + b + f = 180°$ (Angles on a straight line sum to 180°.)
angle e = angle a (Alternate angles are equal.)
angle f = angle c (Alternate angles are equal.)
So $a + b + c = 180°$
Angles inside a triangle always add up to 180°.

Cut any triangle out of a sheet of paper. Colour in the angles at the corners and then tear the triangle into three. You will find that you can place the three angles together along a straight edge (they add up to 180°).

KEYWORDS

Parallel ➤ Parallel lines never meet; they can be marked with arrowheads.

Isosceles triangle ➤ A triangle with two equal sides and two equal angles.

Equilateral triangle ➤ A triangle with all sides and angles equal.

Scalene triangle ➤ A triangle with no equal sides or equal angles.

Right-angled triangle ➤ A triangle with one angle of 90°.

1. Find the missing angles, giving your reasons.

2. Find the angle s.

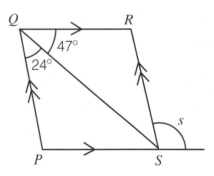

3. Find the angles x, y and z.

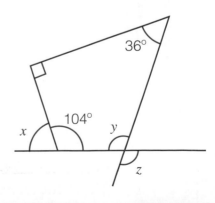

The properties of quadrilaterals

You must know the names of these special quadrilaterals. Quad – four, lateral – side

Name	Sides	Symmetry	
Square	Four sides equal Four angles equal (90°) A **regular** quadrilateral	Four lines Rotational order 4	
Rectangle	Two pairs of opposite sides equal Four angles equal (90°)	Two lines Rotational order 2	
Parallelogram	Two pairs of opposite sides equal and parallel (a squashed rectangle)	No lines Rotational order 2	
Rhombus	Four sides equal Two pairs of opposite sides parallel (a squashed square)	Two lines Rotational order 2	
Trapezium	One pair of parallel sides Could also be isosceles	No lines or one line Rotational order 1	
Kite (also a delta or arrowhead)	Two pairs of adjacent sides equal	One line Rotational order 1	

Make any large polygon shape on the floor – you could use metre rules or just mark the corners with books. Start at one corner and walk round your shape, stopping at each corner and turning. This turn is an external angle. When you get back to your start point turn, facing exactly the same way as you started. You will have turned through 360°. The external angles of a polygon add up to 360°, however many sides it has.

The properties of all polygons

Polygons are named by the number of sides.

Polygons are named by the number of sides.

Sides	3	4	5	6	7	8	9	10
Name	Triangle	Quadrilateral	Pentagon	Hexagon	Heptagon	Octagon	Nonagon	Decagon

A regular polygon has all sides and angles equal. ← A square is a regular quadrilateral.

Polygons have both **external** angles and **internal** angles.
The external angle of any polygon is the angle between one side extended and the next side.
The external angles of any polygon always add up to 360°.
There are two common methods to work out the internal angles in a polygon:

Method 1
➤ Assume the polygon is regular and work out the size of each external angle.
➤ Use the 'angles on a straight line' rule to work out each internal angle.
➤ Calculate the total sum of all the internal angles.

$\frac{360}{5} = 72°$ $180 - 72 = 108°$ $5 \times 108° = 540°$

Method 2
➤ Divide the polygon into triangles.
➤ Use the 'angle sum of a triangle' rule to work out the sum of all the internal angles.
➤ If it is a regular shape, then you can calculate the size of each angle.

$180 \times 3 = 540°$ $\frac{540}{5} = 108°$

1. One angle of an isosceles triangle is 62°. Find the other angles. (Note: there are two possible answers.)
2. Sketch an arrowhead (a special sort of kite), which has three of its angles 20°, 80° and 20°.
3. Calculate the external angle of a regular decagon.
4. If the internal angle of a regular polygon is 175°, how many sides has it got?

Similar and congruent shapes

Shapes are **similar** if they have corresponding angles equal and corresponding sides in the same ratio. Shapes are **congruent** if they have corresponding angles and sides equal.

Congruent shapes are the same shape **and** size.

27

Conditions for similarity

To show two shapes are similar, you must either:

➤ show that all the corresponding pairs of angles are equal, e.g.

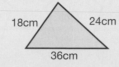

This would mean that one shape is an enlargement of the other.

or

➤ show that all the corresponding pairs of sides are in the same ratio, e.g.

Conditions for congruency

To prove that two triangles are congruent, you must show that one of the following conditions is true.

S – side
A – angle
R – right angle
H – hypotenuse

Side–Side–Side (SSS)	Side, Angle, Side (SAS)	Angle, Corresponding Side, Angle (ASA)	Right angle, Hypotenuse, Side (RHS)
All three sides of one triangle are equal to the three sides in the other triangle.	Two sides and the included angle (the angle between them) of one triangle are equal to the two sides and the included angle in the other triangle.	Two angles and a corresponding side of one triangle are equal to two angles and a corresponding side of the other triangle.	Both triangles have a right angle, an equal hypotenuse and another equal side.
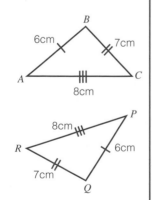	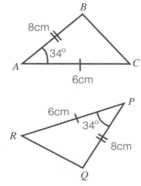	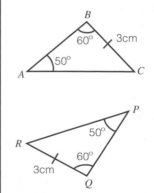	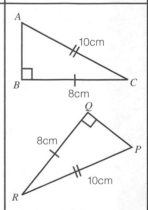
$AB = PQ$, $BC = QR$, $AC = PR$, so congruent (SSS).	$AB = PQ$, $AC = PR$, angle A = angle P, so congruent (SAS).	Angle A = angle P, angle B = angle Q, $BC = QR$, so congruent (ASA).	Angle B = angle Q, $AC = PR$, $BC = QR$, so congruent (RHS).

Congruence and Similarity

Module 27

Proof

You can use congruence to prove angle facts in an isosceles triangle.

An isosceles triangle has two equal sides: $AB = BC$

BD bisects the angle ABC therefore angle ABD = angle CBD

The side BD is common to both triangles.

Therefore SAS: triangle ABD is congruent to triangle CBD

Therefore angle BAD = angle BCD: the base angles are equal.

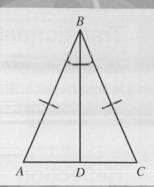

Create a colourful information poster to show the four conditions of congruency. Accurately draw a pair of triangles to illustrate each condition.

KEYWORDS

Similar ➤ Shapes with corresponding angles equal and corresponding sides in the same ratio.

Congruent ➤ Shapes with corresponding angles and sides equal.

1. Rectangle A has sides 3.4m and 5m. Rectangle B has sides 5.1m and 7.5m. Are the rectangles similar?

2. Are these pairs of triangles congruent? Give your reasons if they are congruent.

(a)
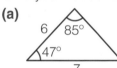

(b)
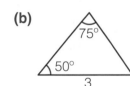 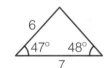

(c)
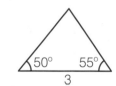

(d)
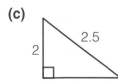 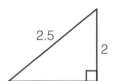

Transformations

All transformations transform an object to make an image. There are four different transformations that you need to know.

Reflection

Every reflection has a **mirror line**. The object is reflected so that the image is reversed and exactly the same distance from the line. ← A mirror image

Remember: To describe a reflection, you need to state the mirror line.

Translation

In a translation, every point slides the same distance across the plane. It is described with a **column vector**.

Remember: To describe a translation, you need to state the vector.

Rotation

Every point in the object is rotated through the same angle about a fixed point.

Remember: To describe a rotation, you need to state the centre of rotation, the angle and the direction.

Enlargement

Enlargement is the only transformation that can give an image which is a different size. The image and the object will always be similar. If the **scale factor** is p, the image of any point will be p times further from the centre of enlargement.

Remember: To describe an enlargement, you need to state the centre of enlargement and the scale factor.

The image and the object are similar shapes.

Translation	Reflection	Rotation	Enlargement
The vector $\begin{pmatrix} 4 \\ -6 \end{pmatrix}$ translates the object O on to the image I, i.e. the object moves +4 in the x–direction and –6 in the y–direction.	A reflection in the line $x = 1$ moves the object O on to the image I.	A rotation of 180° about centre (0, 0) moves the object O on to the image I.	An enlargement of scale factor 2, centre (–2, 0), moves the object O on to the image I.

KEYWORDS

Column vector ➤ A bracket containing values arranged as a column that describe a movement as steps in the x- and the y-directions.

Mirror line ➤ A line used to reflect a shape – it forms a line of symmetry. Points on the mirror line do not move.

Scale factor ➤ The multiplier used to enlarge a shape; the ratio between corresponding sides.

Tricky enlargements

Enlargements can have fractional scale factors:
➤ A scale factor less than 1 gives a smaller image that is closer to the centre of enlargement.

The grid shows O being transformed to I through an enlargement of scale factor $\frac{1}{2}$, centre (0, 0).

Answer the question properly. There is usually a mark for each part of the correct answer.

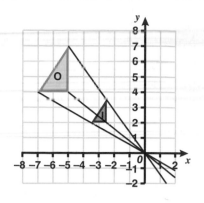

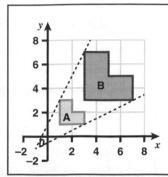

Describe the transformation that moves A on to B.
Enlargement, centre (−1, −1), scale factor 2
or
Describe the transformation that moves B on to A.
Enlargement, centre (−1, −1), scale factor 0.5

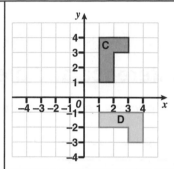

Describe the transformation that moves C on to D.
Rotation, centre (0, 0), 90° clockwise
or
Describe the transformation that moves D on to C.
Rotation, centre (0, 0), 90° anti-clockwise

28

Use a desk light or similar bright light and a pencil to make a shadow on the table. Can you adjust the position of the object (the pencil) to make the image (the shadow) twice as big? Can you get an enlargement of scale factor 3?

1. Describe fully each transformation from green to red.

(a)
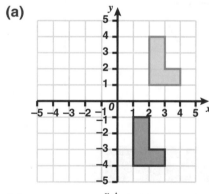

(b)
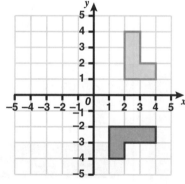

(c)
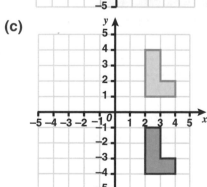

(d)
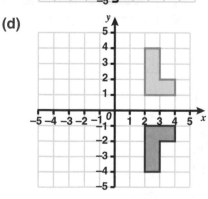

2. What is the **inverse** (opposite) of each of these transformations?
   (a) A reflection in line $y = 2$
   (b) A rotation of 90° clockwise about (2, 1)
   (c) A translation through $\begin{pmatrix} 7 \\ -2 \end{pmatrix}$
   (d) A rotation of 180° about (2, 10)

Circle definitions

You need to know the correct names for the parts of a circle.

The parts of a circle are shown in these diagrams.

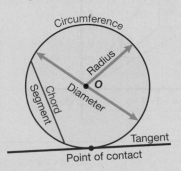

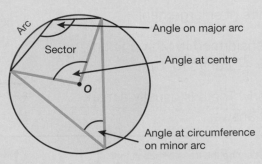

Create an information wall using sheets of A3 paper stuck together. Draw two blank circles and use them to illustrate the different parts of a circle.

The circumference and area of a circle

The perimeter of a circle is called the **circumference**.

You will need to remember both formulae for circumference and area.

> Circumference $C = 2\pi r$ or πd
>
> Area $A = \pi r^2$

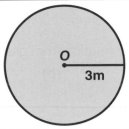

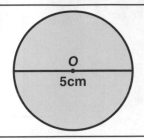

$C = 2\pi r$ or πd $C = 2 \times \pi \times 3$ $\quad = 6 \times \pi = 18.8\text{m}$ (3 s.f.) You can leave your answer 'in terms of π' or you can use your calculator to find a numerical answer.	$C = 2\pi r$ or πd $C = \pi \times 5 = 5\pi$ $\quad = 15.7\text{cm}$ (3 s.f.)
$A = \pi r^2 = \pi \times 3^2 = 9\pi\text{m}^2$	The radius $r = 5 \div 2 = 2.5\text{cm}$ $A = \pi r^2 = \pi \times 2.5^2 = 6.25\pi\text{cm}^2$

Be sure to get the units right:
- Circumference is a length so the units will be mm, cm, m, km …
- Area is measured in squares so the units will be mm², cm², m², km² …

KEYWORDS

Circumference ➤ The length of the perimeter of a circle.

Arc ➤ Part of the circumference.

Sector ➤ Part of a circle formed by two radii and an arc (a 'pie slice').

Arcs and sectors

An **arc** is part of the circumference of a circle. A **sector** is a slice of a circle. You use the angle at the centre to work out the fraction of a circle.

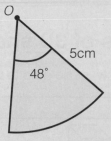

The angle is 48°, which is $\frac{48}{360}$ of the whole circle.

So the arc length is $\frac{48}{360} \times 2\pi \times 5 = \frac{4}{3}\pi$ cm

> You should always state the level of rounding you have used.

$ = 4.19\text{cm (3 s.f.)}$

The perimeter of the whole sector is the arc plus the two radii $= \frac{4}{3}\pi + 5 + 5\text{cm}$ or 14.2cm (3 s.f.)

The area of this sector is $\frac{48}{360} \times \pi \times 5^2 = \frac{10}{3}\pi\,\text{cm}^2$ or 10.5cm² (3 s.f.)

Answering circle questions

Exam questions often include parts of circles. You must be sure what you have to work out. There will be no marks for calculating the area when the question wants the circumference!

A circle question could be on both the non-calculator and calculator papers.

The only difference is how you give your final answer.

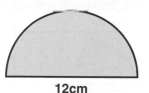

12cm

Non-calculator question	Calculator question
Calculate the area. Leave your answer in terms of π.	Calculate the area. Give your answer to 2 significant figures.
Area $= \frac{\pi r^2}{2} = \frac{\pi \times 6^2}{2}$ $= \frac{36 \times \pi}{2}$ $= 18\pi\,\text{cm}^2$	Area $= \frac{\pi r^2}{2} = \frac{\pi \times 6^2}{2}$ $= \frac{36 \times \pi}{2}$ $= 18\pi$ $= 56.5486... = 57\text{cm}^2$ (2 s.f.)

1. Find the area of a circle with diameter 10cm. Leave your answer in terms of π.

2. Find the perimeter of a semicircle with radius 45cm.

   Give your answer to 3 significant figures.

3. Which has the largest area? A circle of radius 4cm or a semicircle of radius 8cm?

4. Find the perimeter and the area of this shaded sector.

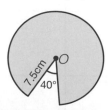

3D shapes

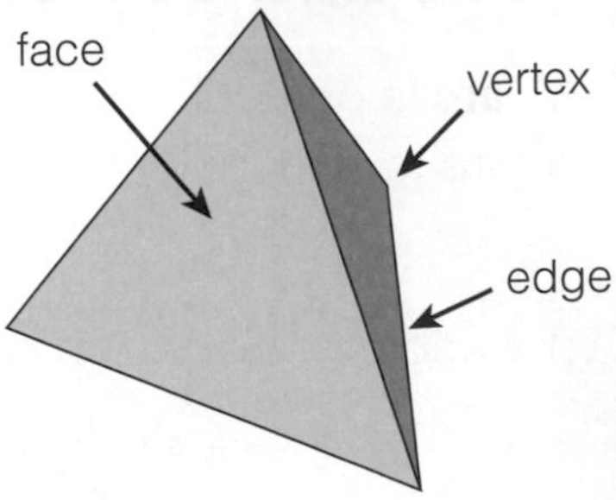

A **3D** shape can be described by its **faces**, **edges** and **vertices** (singular 'vertex').

You may have to work out the area of a face. ← 2D measured in square units like cm²

You may have to work out the volume of the whole shape.

3D measured in cubic units like cm³

face   vertex   edge

Properties of 3D shapes

You need to know, or be able to work out, the properties of these 3D shapes.

	Shape	Name	Vertices	Faces	Edges
Prisms	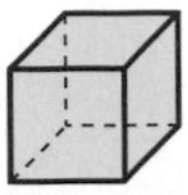	Cube	8	6	12
	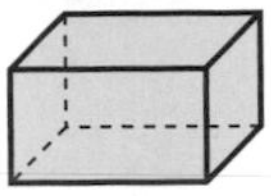	Cuboid	8	6	12
	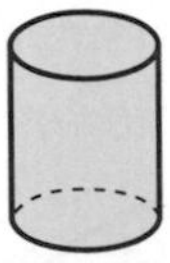	Cylinder	0	3	2
		Triangular prism	6	5	9
	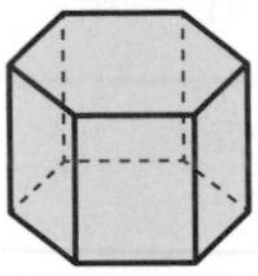	Hexagonal prism	12	8	18
Pyramids		Tetrahedron	4	4	6
	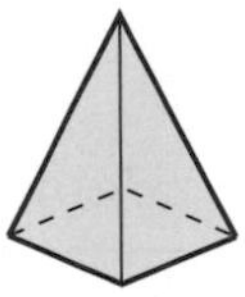	Square-based pyramid	5	5	8
	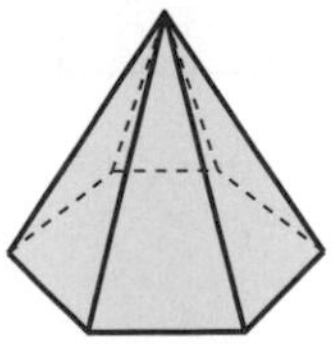	Hexagonal-based pyramid	7	7	12
Curved faces		Cone	1	2	1
		Frustum	0	3	2
	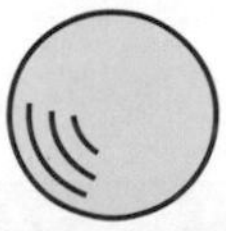	Sphere	0	1	0

Properties of 3D Shapes

Module 30

KEYWORDS

3D ➤ Three dimensions. An object that has length, width and height.

Face ➤ A polygon that makes one surface of a 3D shape.

Edge ➤ Where two faces meet on a 3D shape.

Vertex ➤ The corner of a 2D or 3D shape.

Plan and elevation

Plans and elevations allow accurate 2D representation of 3D shapes.
They are widely used by architects and designers.
The plan is what you would see if you could hover above the shape – it is the outline.
The elevations are what you would see if you stood beside the shape – they are the outlines.

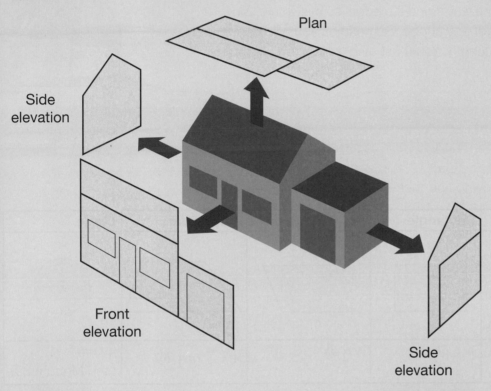

Plan

Side elevation

Front elevation

Side elevation

The Mobius strip – 'every rule has an exception'

Cut a long strip off a sheet of A4 (or larger) stiff paper or card. Twist one short edge by 180° and stick the two short edges together with tape. How many faces and edges has your new shape got? Is it 3D?

1. How many faces, edges and vertices do each of these 3D shapes have?
    (a) A hemisphere
    (b) A pentagonal prism
    (c) A pentagonal-based pyramid

2. Draw the plan and elevations of a hexagonal prism.

3. What shape has a plan and all elevations that are identical circles?

Perimeter, area and volume

Perimeter is a one-dimensional measure. It is a length measured in cm, m, km, inches, miles, etc.

Area is a two-dimensional measure. It is measured in cm², m², km², inches², miles², etc.

Volume is a three-dimensional measure. It is measured in cm³, m³, km³, inches³, miles³, etc.

Perimeter

The perimeter is 'all the way round the edge' so make sure you add up **all** the sides of any shape. Sometimes you will have to calculate a missing length. You must also make sure all the units are the same.

Area

You need to know how to find these areas.

Triangle	Rectangle	Parallelogram	Trapezium	Circle
$A = \frac{1}{2}\,bh$	$A = bh$	$A = bh$	$A = \frac{1}{2}\,(a+b)h$	$A = \pi r^2$ (see Module 29)

h is the vertical height, at right angles to the base.

You must make sure all the units are the same and write the correct 'square unit' with your answer.

Find areas of **composite shapes** by chopping them up into any of the simple shapes above.

The two area formulae below are given in your exam, so you will not have to learn them, but you must understand them and be able to use them correctly:

> **Curved surface area of a cone** = $\pi r l$ (where l is the slant height of the cone)
>
> **Surface area of a sphere** = $4\pi r^2$

A sphere of radius 3cm has surface area $4\pi \times 3^2 = 113\text{cm}^2$ (3 s.f.)

Volume

You need to know how to find these volumes.

Any prism	Any pyramid
Area of the cross-section × height (or length)	$\frac{1}{3}$ area of the cross-section × height

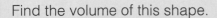

Volume = 3 × 4 × 7 = 84cm³

Volume = $\frac{1}{3}$ × 3 × 4 × 7 = 28cm³

Find more complicated volumes by chopping them up into any of the simple volumes above.

Find the volume of this shape.

Volume of prism = area of cross-section × length

Area of cross-section = 10 × 4 + 4 × 4 ←

= 56cm²

Split into a rectangle and a square.

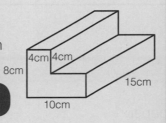

Volume = 56 × 15 = 840cm³

These two volume formulae are given in your exam, so you will not have to learn them, but you must understand them and be able to use them correctly:

Volume of a sphere = $\frac{4}{3}\pi r^3$	**Volume of a cone** = $\frac{1}{3}\pi r^2 h$

Make three larger copies of this net. Fold them to make three identical pyramids that will fit exactly together to make a cube.

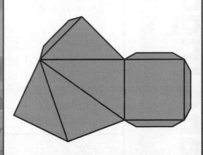

---

1. Find the perimeter and area of these shapes.

   (a)

   10m
   4m
   4m
   3m
   3m

   (b)

   7cm
   5cm
   12cm
   8cm
   10cm

2. Find the volume of these shapes.

   (a)

   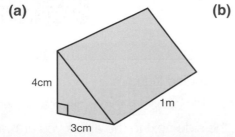

   4cm
   1m
   3cm

   (b)

   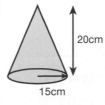

   20cm
   15cm

   (c)

   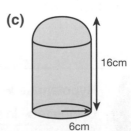

   16cm
   6cm

Pythagoras' theorem

Use Pythagoras' theorem for finding the missing side in a right-angled triangle when you have been given the other two lengths.

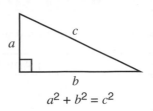

$$a^2 + b^2 = c^2$$

The easiest way to remember Pythagoras is using a diagram.

Remember that the squares of the two **short** sides add to give the square of the **longest** side.

Calculate AC.	Calculate PQ.
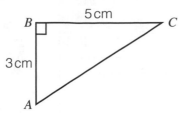 5 cm, 3 cm, B, C, A	P, 12 m, 9 m, Q, R
$a^2 + b^2 = c^2$	$a^2 + b^2 = c^2$
$3^2 + 5^2 = AC^2$	$PQ^2 + 9^2 = 12^2$
$AC^2 = 34$	$PQ^2 = 144 - 81 = 63$
$AC = \sqrt{34}$ cm	$PQ = \sqrt{63}$ m
➤ Non-calculator paper: Leave answer as a square root.	➤ Non-calculator paper: Leave answer as a square root.
➤ Calculator paper: Give rounded decimal answer.	➤ Calculator paper: Give rounded decimal answer.
$AC = \sqrt{34} = 5.83$ cm (3 s.f.)	$PQ = \sqrt{63} = 7.94$ m (3 s.f.)

Basic trigonometry ratios

Basic trigonometry ratios only work for right-angled triangles. They are used to find angles from sides and sides from an angle and a side. Every angle has its own set of sin/cos/tan values.

$$\mathbf{s}\text{ine}\,\theta = \frac{\mathbf{opposite}}{\mathbf{hypotenuse}} \left(s = \frac{o}{h}\right) \quad \mathbf{c}\text{osine}\,\theta = \frac{\mathbf{adjacent}}{\mathbf{hypotenuse}} \left(c = \frac{a}{h}\right) \quad \mathbf{t}\text{angent}\,\theta = \frac{\mathbf{opposite}}{\mathbf{adjacent}} \left(t = \frac{o}{a}\right)$$

> You could learn a mnemonic to help you remember them. Can you make up a better mnemonic than 'SOHCAHTOA'? Silly Old Henry Can…

You need to practise with your calculator.

To solve $a = \sin 45$: key [sin] [4] [5] [=] to get the length $a = 0.707…$

> You may have to close a bracket before hitting [=]

To solve $\sin a = 0.6$: key [sin⁻¹] [0] [.] [6] [=] to get the angle
$a = 36.86…°$

> You may have to use the [SHIFT] button to get sin⁻¹.

KEYWORDS

Hypotenuse ➤ The longest side in a right-angled triangle.

Opposite ➤ In a right-angled triangle, the side opposite the angle you are working with.

Adjacent ➤ In a right-angled triangle, the side next to the angle you are working with.

Using trigonometry

Check you can get the correct answers from your calculator for these examples:

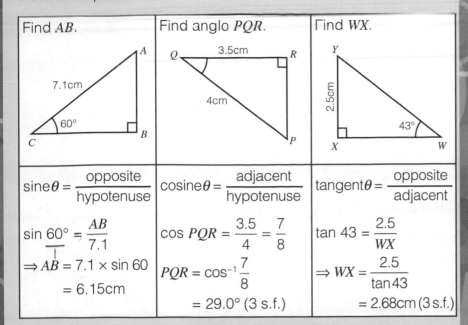

Find AB.	Find angle PQR.	Find WX.
$\sin\theta = \dfrac{\text{opposite}}{\text{hypotenuse}}$	$\cos\theta = \dfrac{\text{adjacent}}{\text{hypotenuse}}$	$\tan\theta = \dfrac{\text{opposite}}{\text{adjacent}}$

$\sin 60° = \dfrac{AB}{7.1}$

$\Rightarrow AB = 7.1 \times \sin 60$

$= 6.15\text{cm}$

$\cos PQR = \dfrac{3.5}{4} = \dfrac{7}{8}$

$PQR = \cos^{-1}\dfrac{7}{8}$

$= 29.0°$ (3 s.f.)

$\tan 43 = \dfrac{2.5}{WX}$

$\Rightarrow WX = \dfrac{2.5}{\tan 43}$

$= 2.68\text{cm}$ (3 s.f.)

Using exact values for sin, cos and tan

Your calculator will give you decimal values of sin, cos and tan but you can also work out exact values for some special angles.

Angle	sin	cos	tan
0°	0	1	0
30°	$\dfrac{1}{2}$	$\dfrac{\sqrt{3}}{2}$	$\dfrac{1}{\sqrt{3}}$
60°	$\dfrac{\sqrt{3}}{2}$	$\dfrac{1}{2}$	$\sqrt{3}$
45°	$\dfrac{1}{\sqrt{2}}$	$\dfrac{1}{\sqrt{2}}$	1
90°	1	0	Does not exist

Draw any right-angled triangle accurately. Draw a square on each side like this. Cut out the two smaller squares. Can you cut them up to fit exactly inside the large square?

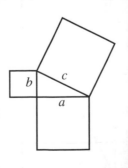

1. Is a triangle with sides 4.2cm, 7cm and 5.6cm right-angled? Explain your answer.

2. Calculate the missing angles.

   (a)
   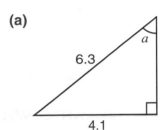

   (b)
   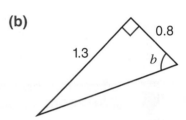

   (c)
   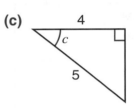

3. Calculate the missing lengths.

   (a)
   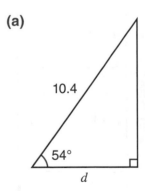

   (b)
   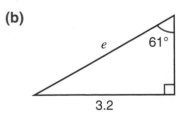

   (c)
   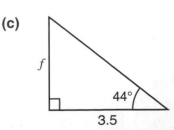

Vectors and translations

A vector can be written as a column vector, as a single bold letter plus an arrow, or with letters and an arrow on the top. It has both length and direction.

On this graph (2, 1) is the fixed point, A.

The column vector $\begin{pmatrix} 1 \\ 3 \end{pmatrix}$ is a **translation** of any point +1 in the x-direction and +3 in the y-direction. On this graph it is the movement from A to C

and can also be written as **a** (with an arrow to show direction) or $\overrightarrow{AC}$.

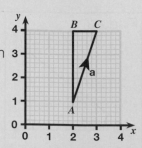

Working with vectors

You can add and subtract vectors.
Think of this as one translation
followed by another.

You can multiply any vector by a **scalar**
which makes it larger or smaller.

a		$\begin{pmatrix} 1 \\ 2 \end{pmatrix}$
b		$\begin{pmatrix} 2 \\ -3 \end{pmatrix}$
a + b		$\begin{pmatrix} 1 \\ 2 \end{pmatrix} + \begin{pmatrix} 2 \\ -3 \end{pmatrix} = \begin{pmatrix} 3 \\ -1 \end{pmatrix}$
a – b		$\begin{pmatrix} 1 \\ 2 \end{pmatrix} - \begin{pmatrix} 2 \\ -3 \end{pmatrix} = \begin{pmatrix} -1 \\ 5 \end{pmatrix}$
2a		$\begin{pmatrix} 2 \\ 4 \end{pmatrix}$
$\frac{1}{2}$ **b**		$\begin{pmatrix} 1 \\ -1.5 \end{pmatrix}$
2a – b		$\begin{pmatrix} 2 \\ 4 \end{pmatrix} - \begin{pmatrix} 2 \\ -3 \end{pmatrix} = \begin{pmatrix} 0 \\ 7 \end{pmatrix}$

More about vectors

The **inverse** of a vector $\begin{pmatrix} u \\ b \end{pmatrix}$ is the vector $\begin{pmatrix} -a \\ -b \end{pmatrix}$

If you have to describe a vector 'in terms of' other vectors, just find your way between the points using the vectors you already have.

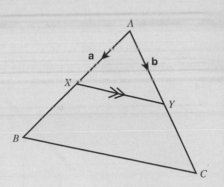

X and Y are the midpoints of AB and AC respectively.

$\overrightarrow{XY} = -\mathbf{a} + \mathbf{b}$ or $\mathbf{b} - \mathbf{a}$ ← Vectors obey the same rules as numbers.

$\overrightarrow{YX} = -\mathbf{b} + \mathbf{a}$ or $\mathbf{a} - \mathbf{b}$

$\overrightarrow{AB} = 2\mathbf{a}$ $\qquad$ $\overrightarrow{BA} = -2\mathbf{a}$

$\overrightarrow{AC} = 2\mathbf{b}$ $\qquad$ $\overrightarrow{CA} = -2\mathbf{b}$

$\overrightarrow{BC} = -2\mathbf{a} + 2\mathbf{b}$

Use vectors to describe your movement around a room. You will need to decide which is the x-direction and which is the y-direction. For example, $\begin{pmatrix} 3 \\ 4 \end{pmatrix}$ could be three steps to the right and four steps forwards.

---

1. If $\mathbf{p} = \begin{pmatrix} 7 \\ -1 \end{pmatrix}$, $\mathbf{q} = \begin{pmatrix} 2 \\ 2 \end{pmatrix}$ and $\mathbf{r} = \begin{pmatrix} 0 \\ -1 \end{pmatrix}$ calculate:

   (a) $\mathbf{p} + \mathbf{r}$ $\qquad$ (b) $\mathbf{p} + \mathbf{q} - \mathbf{r}$

   (c) $3\mathbf{p} + 2\mathbf{q}$ $\qquad$ (d) $\mathbf{p} - 2\mathbf{r}$

2. $OABC$ is a quadrilateral, $\overrightarrow{OA} = \mathbf{a}$, $\overrightarrow{OB} = \mathbf{b}$ and $\overrightarrow{OC} = \mathbf{c}$. P, Q, R and S are the midpoints of OA, AB, BC and OC respectively.

   Write $\overrightarrow{OP}$, $\overrightarrow{AB}$ and $\overrightarrow{CR}$, in terms of $\mathbf{a}$, $\mathbf{b}$ and $\mathbf{c}$.

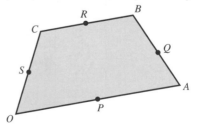

33

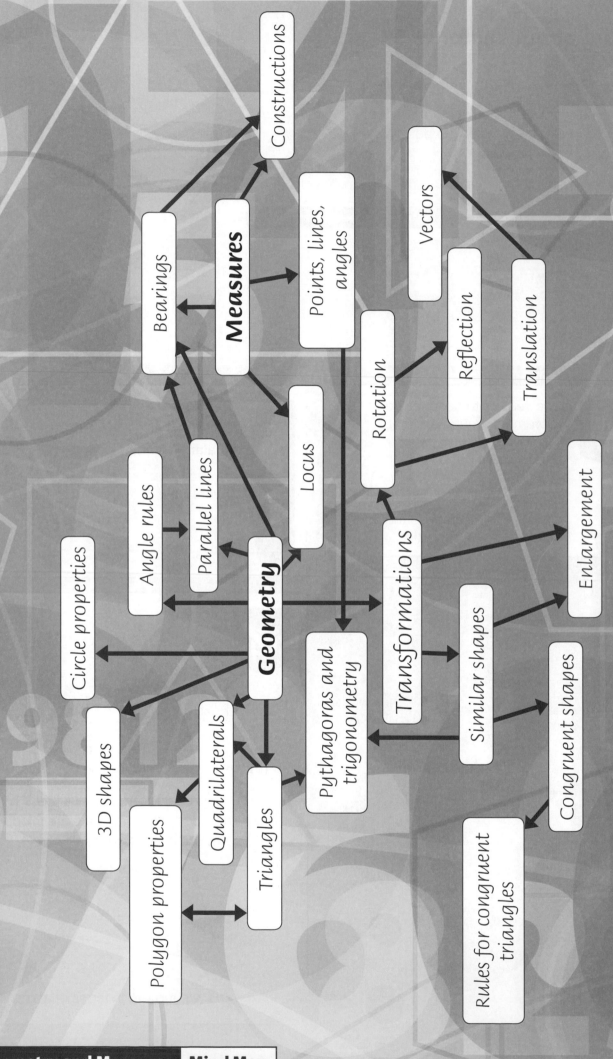

1. If the interior angle of a regular polygon is 156°, how many sides does it have? **[2]**

2. Find the angle *HGD* giving all your reasons. 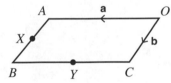 **[3]**

3. In triangle *ABC* the length *AB* = 10cm, angle *BAC* = 92° and angle *ABC* = 20°.
In triangle *RST* the length *SR* = 10cm, angle *RST* = 20° and angle *STR* = 68°.

Draw triangles *ABC* and *RST* and show that they are congruent, giving your reasons. **[3]**

4. (a) Rotate shape A 90° clockwise about the point (0, 0) and label the image B. **[2]**

(b) Reflect B in the line $x = 0$ and label the image C. **[1]**

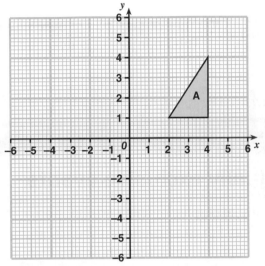

5. *OABC* is a parallelogram with *X* the midpoint of *AB* and *Y* the midpoint of *BC*.

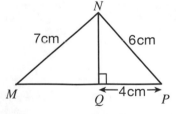

Use the vectors **a** and **b** to write:

(a) $\overrightarrow{OA}$ **[1]**   **(b)** $\overrightarrow{OB}$ **[1]**   **(c)** $\overrightarrow{OY}$ **[2]**

6.

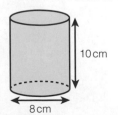

(a) Find the length *NQ*. Give your answer to 2 decimal places. **[3]**

(b) Find the angle *NMQ*. Give your answer correct to 3 significant figures. **[3]**

7. Find the volume of this cylinder. Give your answer in terms of π. 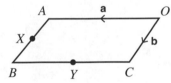 **[3]**

Theoretical probability

Probabilities can be written as fractions, decimals or percentages.
The probability of an event must be between 0 and 1 on the probability scale.

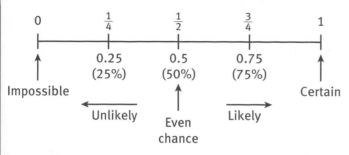

0    $\frac{1}{4}$ $\frac{1}{2}$ $\frac{3}{4}$ 1

0.25   0.5   0.75
(25%)   (50%)   (75%)

Impossible   Certain

← Unlikely →   Likely →

Even
chance

To find the probability of an event, use the formula:

$$P(\text{event}) = \frac{\textbf{Number of successful outcomes}}{\textbf{Total number of outcomes}}$$

$P(\text{drawing a king from a deck of cards}) = \frac{4}{52}$ ← 4 kings in a deck
← 52 cards in a deck

Mutually exclusive events

Mutually exclusive events can't happen at the same time. For example, when you roll a dice, you can't get a 1 and a 6 at the same time.

The probabilities of all the outcomes of an event must add up to 1.

The probability of getting each number on a dice is $\frac{1}{6}$ and the probabilities all add up to 1, since it is certain that one of them will occur.

If the probability of event A happening is P, then the probability of event A **not** happening is 1 – P.
P(A) = P   P(not A) = 1 – P

If the probability of a white Christmas is 0.23, then the probability of not having a white Christmas is

1 – 0.23 = 0.77 ← The total probability of the two events must equal 1.

KEYWORDS

P(event) ➤ Probability of an event happening.

Mutually exclusive ➤ Events that can't happen at the same time.

Frequency ➤ The number of times something happens.

Biased ➤ Unfair, i.e. weighted on one side.

34

Experimental and Theoretical Probability

Module 34

Relative frequency

It is not always possible to calculate a theoretical probability. In such cases the probability can be estimated by doing experiments or trials – this is called relative **frequency**.

This **biased** spinner is spun 20 times and then 100 times. Here are the results:

Score	1	2	3	4	Total trials
First test	2	3	1	14	20
Second test	11	14	7	68	100

$P(4) = \dfrac{14}{20} = 0.7$

$P(4) = \dfrac{68}{100} = 0.68$

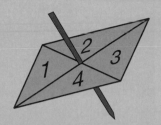

The more trials that are done, the more accurate the estimated probability will be. 0.68 is a better estimate than 0.7 for the probability of getting a 4 on this biased spinner.

Expected probability

If the probability of an event is known, then it is easy to predict the number of times that the event is expected to happen. Use the formula:

Expected number of occurrences = Probability of event × Number of trials

The probability that a biased coin will land on heads is 0.24. If the coin is tossed 600 times, how many heads will be expected to occur?

Expected number of heads = 0.24 × 600 = 144

1. A fair dice is rolled. Find:
   (a) P(even) **(b)** P(prime) **(c)** P(factor of 6)
2. The probability that a person will pass their driving test is 0.46
   (a) Find the probability that a person will not pass their driving test.
   (b) 500 people take their test this month. How many do you expect to pass?
3. Mary and Max both do an experiment to find out whether a drawing pin will land point up. Mary drops the pin 20 times and Max drops the pin 50 times. Here are the results:

	Point up	Total trials
Mary	13	20
Max	34	50

   (a) Who has the most reliable results and why?
   (b) What is the best estimate for the probability that the pin will land point up?

Roll a dice 30 times and record your results. Do the results show what you would have expected?
Do you believe your dice is biased or fair?

Roll the dice again 60 times and record your results. Are the results now closer to what you would expect?

Sample space

A **sample space** diagram is a list or a table which shows all the possible outcomes of two or more **combined** events.

Two dice are rolled and the scores are added together:

6	7	8	⑨	10	11	12
5	6	7	8	⑨	10	11
4	5	6	7	8	⑨	10
3	4	5	6	7	8	⑨
2	3	4	5	6	7	8
1	2	3	4	5	6	7
	1	**2**	**3**	**4**	**5**	**6**

Dice 1 (vertical axis) — Dice 2 (horizontal axis)

The probability of scoring 9 is $\frac{4}{36}$

A meal deal offers a choice of sandwiches plus a snack.

You can choose either a ham (H), cheese (C) or egg (E) sandwich **and** either chocolate (ch) or crisps (cr).

Find the probability that someone at random chooses an egg sandwich and crisps.

You can write all of the outcomes as a list:

(H, ch), (H, cr), (C, ch), (C, cr), (E, ch), (E, cr)

There are exactly six outcomes. Do not write down any repeat outcomes, e.g. (ch, H).

P(egg and crisps) = $\frac{1}{6}$

Independent events

Independent events can happen at the same time and the occurrence of one event does not affect the occurrence of the other. For example, if you roll a dice and flip a coin, you can get a head and a six and one event does not affect the other.

Learn the following rules:

Rule	What you say	What you do	
AND rule	P(A and B)	P(A) × P(B) ←	**AND** means **times**.
OR rule	P(A or B)	P(A) + P(B) ←	**OR** means **add**.

Find the probability of getting a head on a coin and a six on a dice.

P(head AND six) = P(head) × P(six)

$$= \frac{1}{2} \times \frac{1}{6} = \frac{1}{12}$$

Tree diagrams

A tree diagram is a useful way of showing all of the outcomes of two or more combined events. They are also an excellent tool for calculating probabilities.

A coin is tossed and the spinner is spun.

Find the probability of getting a head on the coin and an odd number on the spinner.

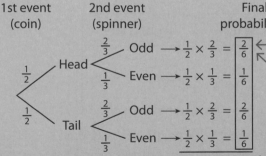

1st event (coin) — 2nd event (spinner) — Final probabilities

Head: $\frac{1}{2}$
- Odd $\frac{2}{3}$ → $\frac{1}{2} \times \frac{2}{3} = \frac{2}{6}$
- Even $\frac{1}{3}$ → $\frac{1}{2} \times \frac{1}{3} = \frac{1}{6}$

Tail: $\frac{1}{2}$
- Odd $\frac{2}{3}$ → $\frac{1}{2} \times \frac{2}{3} = \frac{2}{6}$
- Even $\frac{1}{3}$ → $\frac{1}{2} \times \frac{1}{3} = \frac{1}{6}$

Total = 1

Multiply across the branches (AND rule).

P(head and odd) = $\frac{2}{6}$

The four different outcomes must add up to 1 (OR rule).

Venn diagrams and set notation

Venn diagrams can also be used to represent and to calculate probabilities.

Some students took a music practical exam and a music theory exam. Everyone passed at least one exam. 46% passed the practical exam and 84% passed the theory exam. Find the probability that a student chosen at random passed the practical exam only.

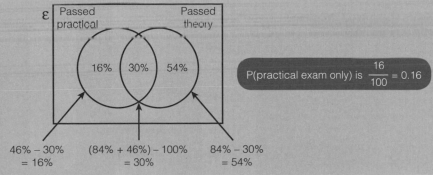

$P(\text{practical exam only})$ is $\dfrac{16}{100} = 0.16$

| $46\% - 30\%$ | $(84\% + 46\%) - 100\%$ | $84\% - 30\%$ |
| $= 16\%$ | $= 30\%$ | $= 54\%$ |

This table shows some set notation which is represented by the Venn diagram below.

Set notation	Meaning	Set of numbers
$\mathcal{E} = \{x: 0 < x < 10\}$	Universal set = all items that could occur	{1, 2, 3, 4, 5, 6, 7, 8, 9}
$A = \{x: 0 < x < 10 \text{ and } x \text{ is odd}\}$	Set A = odd numbers from 0 to 10	Set A = {1, 3, 5, 7 and 9}
$B = \{x: 0 < x < 10 \text{ and } x \text{ is prime}\}$	Set B = prime numbers from 0 to 10	Set B = {2, 3, 5 and 7}

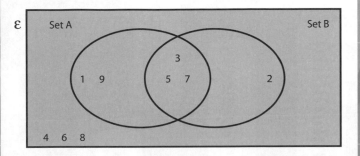

1. Two dice are rolled and the scores are added together. Using the sample space diagram on page 80, find:
   (a) P(double 6) (b) P(not getting a double)
   (c) What is the most likely score to occur?
2. There are 2 black and 3 white counters in a bag. One counter is taken at random, the colour noted and then replaced. A second counter is then taken. Find the probability that both of the counters are black.

Flip a coin three times. Repeat the experiment 40 times. Record the number of times that you get three heads in a row.

Draw a tree diagram for three coins and compare your theoretical results with your experimental results. How close are they?

Probability

- **Theoretical**
  - Probability scale
  - Fractions, decimals and percentages
  - Mutually exclusive
    - Add up to 1
    - $P(A) = 1 - P(\text{not } A)$
  - Combined events
    - Independent events
      - AND/OR rule
    - Set notation
  - Equally likely outcomes
    - Random
    - Fair
      - Bias
        - Interpret

- **Experimental**
  - Estimated
  - Sample size
    - Interpret
    - Bias
  - Relative frequency
    - Two-way tables

- **Representing**
  - Venn diagrams
    - Set notation
  - Tree diagrams
    - Venn diagrams
  - Sample space
    - Tree diagrams
  - Two-way tables

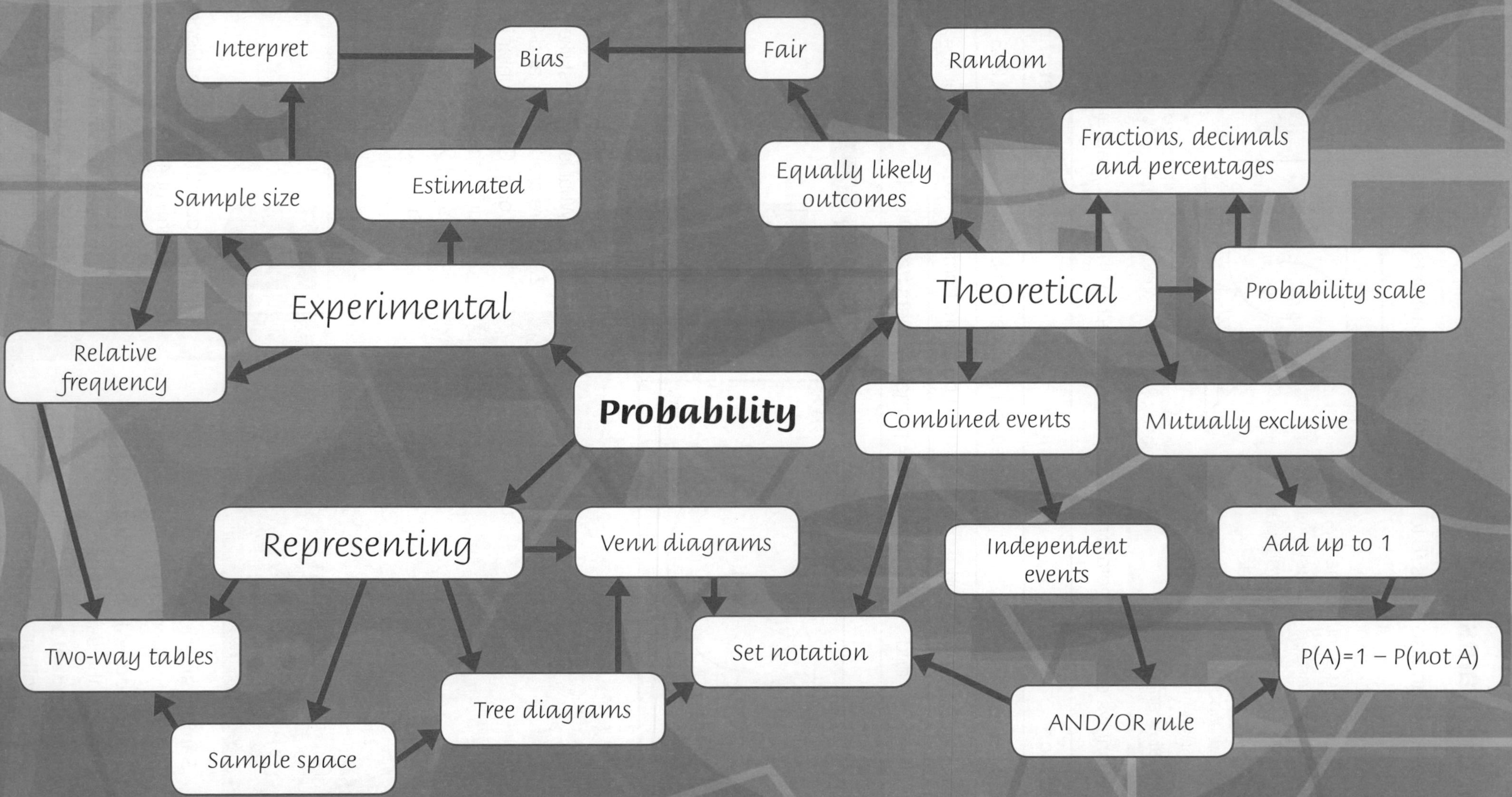

1. A letter is chosen at random from the word PROBABILITY. Find the probability that the letter will be: Ⓕ

   (a) A letter B **[1]**   (b) A vowel **[1]**

   (c) Not an I **[1]**   (d) A letter Q **[1]**

2. The probability that a biased dice will land on the numbers 1 to 5 is given in the table.

Number	1	2	3	4	5	6
Probability	0.1	0.2	0.25	0.1	0.15	

   (a) Find the probability that the dice will land on 6.   **[2]**

   (b) The dice is rolled 60 times. How many times do you expect it to land on 5?   **[2]**

3. Two fair coins are flipped. Find the probability of getting: Ⓕ

   (a) Two heads   **[2]**

   (b) At least one tail   **[2]**

4. A cinema has a £6 food and drink deal. You may choose either a hotdog (D), nachos (N) or popcorn (P) **and** either tea (T), coffee (C), hot chocolate (H) or a soft drink (S).

   (a) Write a list of all the possible outcomes. The first one is done for you:
      (D, T) …   **[2]**

   (b) Find the probability that someone at random chooses a hotdog and a coffee.   **[2]**

5. Mark and Simon play snooker and then table tennis.
   The probability that Mark will win at snooker is 0.3
   The probability that Mark will win at table tennis is 0.6

   (a) Copy and complete the tree diagram.   **[2]**

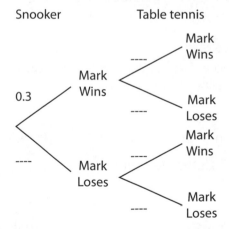

   (b) Find the probability that Mark will win both games.   **[2]**

   (c) Find the probability that Mark will win at least one game.   **[2]**

6. 50 students were asked to try chocolate A and chocolate B.

   Here are the results: Ⓕ

   45 said they liked chocolate A   46 said they liked chocolate B   42 said they liked both

   (a) Draw a Venn diagram to show the information.   **[2]**

   (b) Find the probability that a student chosen at random did not like either.   **[2]**

   (c) Find the probability that a student chosen at random only liked chocolate A.   **[1]**

Types of data

You need to know the difference between different types of data.

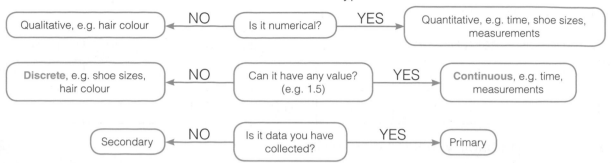

| Qualitative, e.g. hair colour | ← NO — | Is it numerical? | — YES → | Quantitative, e.g. time, shoe sizes, measurements |

| **Discrete**, e.g. shoe sizes, hair colour | ← NO — | Can it have any value? (e.g. 1.5) | — YES → | **Continuous**, e.g. time, measurements |

| Secondary | ← NO — | Is it data you have collected? | — YES → | Primary |

Averages

There are three types of average – **mean**, **median** and **mode**. These are used to help compare different **populations**.

36

Ungrouped data

The table below shows the goals scored by a school football team in 70 games.

Goals scored	Frequency	Total goals	
0	22	0 × 22 =	0
1	21	1 × 21 =	21
2	13	2 × 13 =	26
3	10	3 × 10 =	30
4	2	4 × 2 =	8
5	2	5 × 2 =	10
	70		95

↑ Total number of games

↑ Total number of goals scored in all the 70 games

Mean = $\dfrac{\text{Total goals scored}}{\text{How many games}}$

= 95 ÷ 70 = 1.4 goals per game

Median interval: The group with the middle value. So out of 70 games, halfway is between the 35th and 36th value. Both of these will be in the second group. So the median value is 1.

Mode: Most common number of goals scored is 0 (the group with the highest frequency). So the mode is 0.

Grouped data

The table below shows the times taken by 200 runners in a race.

Time (mins)	Frequency	Midpoint	Frequency × Midpoint
10 ≤ t < 15	1	12.5	12.5
15 ≤ t < 25	8	20	160
25 ≤ t < 35	33	30	990
35 ≤ t < 45	56	40	2240
45 ≤ t < 55	48	50	2400
55 ≤ t < 65	35	60	2100
65 ≤ t < 75	19	70	1330
	200		9232.5

↑ Total number of runners

↑ This is the assumed running time for every person in that group.

↑ Total time taken by all the runners

Estimated mean

= $\dfrac{\text{Total time for all runners}}{\text{How many runners}}$

= 9232.5 ÷ 200 = 46.2 minutes

↑ This is only an estimated mean as it uses the midpoint for each group and not the actual values of the times for each person.

Median interval: Between 100th and 101st runner – both these values will be in the fifth group. So the median will be in the group 45 ≤ t < 55.

Modal group: 35 ≤ t < 45 (the group with the highest frequency)

Module 36 Data and Averages

Sampling

To find out about a population or to test a **hypothesis**, a sample is taken and analysed.

How has it been collected? Does anything make it biased? ← What makes a good sample? → Sample size 3 to 10% of the population

There are different methods of picking a sample (e.g. systematic – pick every tenth one or do it randomly by using a random number generator) and there are different sample types (e.g. stratified – the sample has the same proportion of each group as per the population).

For example, a supermarket chain wants to find out how much people spend on groceries in a week. It asks the first 30 people who shop at its Watford branch in December. This sample would be **biased** as it is limited to one store and so is not a **representative sample**. Also, 30 people is not a large enough sample.

Conduct your own survey on how many text messages people send in a week. Do this for both boys and girls separately. Use a table like this to record your data.

Number of texts	Boys			Girls		
	Freq.	Midpt.	F×M	Freq.	Midpt.	F×M
$0 \leqslant t < 20$						
$20 \leqslant t < 40$						
$40 \leqslant t < 60$						
$60 \leqslant t < 80$						
$80 \leqslant t < 100$						
$100 \leqslant t < 150$						
$150 \leqslant t < 200$						

Create a short video clip on how you calculate the mean for the boys and the mean for the girls.

Two-way tables

Two-way tables show two different variables. Examples include train and bus timetables.

Watford	07.35	08.05	08.35	09.05	10.05
Birmingham	09.25	09.55	10.25	10.55	11.55
Sheffield	10.23	10.53	11.23	11.53	12.23
Leeds	11.36	12.06	12.36	13.06	14.06
Sunderland	12.42	13.12	13.42	14.12	15.12
Newcastle	13.48	14.18	14.48	15.18	16.18

The columns represent the different trains. For example, there is a train leaving Watford at 10.05am and arriving in Leeds at 2.06pm.

KEYWORDS

Discrete data ➤ Can only have set values, e.g. hair colour, shoe size, trouser size.

Continuous data ➤ Can take any value, e.g. things you measure (time, length, weight).

Mean ➤ Total of values divided by the number of values.

Median ➤ The middle value when the data is ordered.

Mode ➤ The most common value.

Population ➤ The data set, e.g. people or objects.

Sampling ➤ A way of collecting data from a population.

Hypothesis ➤ An idea you believe to be true.

1. When working with grouped data, why can you only find an estimate of the mean?

2. Adam has a hypothesis that girls spend more time surfing the Internet than boys. He takes a sample of 10 girls and 10 boys in his class. Is this a good sample? Why?

3. Look at the train timetable (left). Ruth needs to be in Newcastle for an interview at 3.00pm. It will take her 25 minutes to get to the interview from the train station once she arrives. She lives in Sheffield. What is the latest train she can catch and how long will it take?

Pictograms, Venn diagrams, bar charts and line graphs

Pictograms, Venn diagrams, bar charts and vertical line graphs are best for discrete and qualitative data. As well as being able to draw them, you need to be able to read them and pick an appropriate diagram for the situation.

The **pictogram** shows the results of a survey on the different cereals eaten at a school breakfast club.

Venn diagrams are used to show relationships between sets of data. For example, in the diagram to the right the first set is the number of students who study Geography and the second set is the number of students who study History. The overlap represents the students who study both subjects and the space outside the circles represents students who study neither subject.

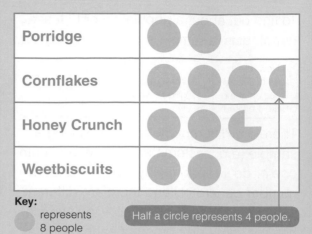

Porridge	
Cornflakes	
Honey Crunch	
Weetbiscuits	

Key:

represents 8 people

Half a circle represents 4 people.

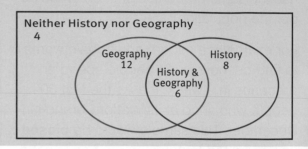

Neither History nor Geography
4

Geography 12

History & Geography 6

History 8

The number of complaints at a call centre was recorded over a period of two weeks. **Bar charts** and **line graphs** can be used to compare this data.

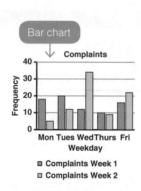

Bar chart

	Complaints Week 1	Complaints Week 2
Mon	18	5
Tues	20	12
Wed	12	34
Thurs	10	9
Fri	16	22

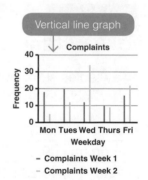

Vertical line graph

Pie charts

Pie charts work best for discrete data. Here are the results of a survey of 60 students from Year 7 on how many text messages they send in a week.

This is the number of people in each category.

Number of texts	Frequency	Angle
$0 \leqslant t < 50$	18	$360° \div 60 \times 18 = 108°$
$50 \leqslant t < 100$	20	$360° \div 60 \times 20 = 120°$
$100 \leqslant t < 200$	12	$360° \div 60 \times 12 = 72°$
$200 \leqslant t < 300$	10	$360° \div 60 \times 10 = 60°$

Remember that there are 360° in a circle. So divide 360 by the total frequency (60 students), then multiply by the frequency for each category.

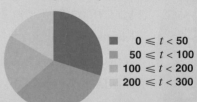

Survey of Year 7 students – text messages sent per week

- $0 \leqslant t < 50$
- $50 \leqslant t < 100$
- $100 \leqslant t < 200$
- $200 \leqslant t < 300$

Line graphs for time series data

Some types of data that are measured against time can be represented on a time series graph. Examples include prices, Retail Price Index, share prices, sales, etc. You will need to be able to plot and interpret these.

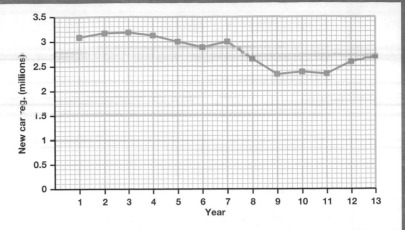

Scatter graphs

Scatter graphs are used to test for a relationship between two variables. They will show positive **correlation**, negative correlation or no correlation.

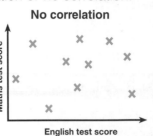

Positive correlation   **Negative correlation**   **No correlation**

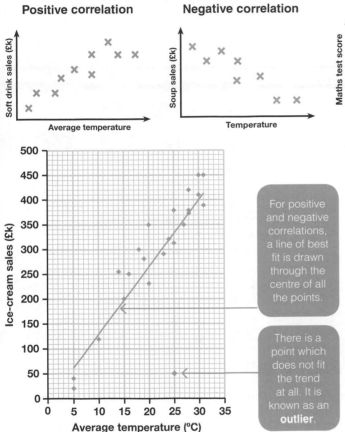

For positive and negative correlations, a line of best fit is drawn through the centre of all the points.

There is a point which does not fit the trend at all. It is known as an **outlier**.

A correlation can be weak or strong, depending on how closely the points follow the line of best fit. You can use a line of best fit to make predictions but only **within the range of the data**. It would be unreliable to extend the line of best fit and make predictions beyond the range of the data because the relationship may not hold.

Create your own pie chart on an A4 sheet of paper to show the number of pages in 40 books in your home. You might group the books into categories of 0–50 pages, 51–100 pages, 101–200 pages, 201–300 pages, 301 or more pages, etc.

37

1. Look at the time series graph above.
   (a) What could have caused the dip in years 7–9 in new car registrations?
   (b) How many new car registrations would you predict for year 14?

Comparing data

When comparing data you will be expected to be able to interpret any of the diagrams in Module 37, as well as use the averages and the range to compare distributions.

Measures of central tendency – the averages

Consider the scores below for a maths test:

23   10   34   21   41   39   23   34   30   22   34   32

Putting this data in order: 10, 21, 22, 23, 23, 30, 32, 34, 34, 34, 39, 41

> The mean will lie somewhere within the range of this data, so it will not be less than 10 and it will be no more than 41.

Mean: 343 ÷ 12 = 28.6

> Total of test scores

> Number of scores in the data

Mode: 34 ← The test score that occurs most times.

Median: 31 ← The score that represents the middle value when the data is ordered. In this case, the median lies between the sixth and seventh values because there is an even number of test scores in the data.

Range: 41 – 10 = 31 ← Maximum test score – Minimum test score

Consider what happens to the averages when you add an outlier to the data:

10, 21, 22, 23, 23, 30, 32, 34, 34, 34, 39, 41, **95**

Mean is now 438 ÷ 13 = 33.7 ← Higher than before   Mode is 34 ← No change

Median is now the seventh value = 32 ← Small change   Range is now 95 – 10 = 85 ← Big change

Average	Advantage	Disadvantage
Mean	Uses every value so represents all the data	Affected by outliers (extreme values)
Median	Not affected by outliers	Only represents a single value in the middle of the distribution
Mode	Used for numbers and words; not affected by outliers	Limited use
Range	Gives a guide on how spread out the distribution is	Only uses minimum and maximum values and can be highly affected by outliers

Comparing distributions from a list

In a survey of gym members, the time (in hours) spent exercising in a week was recorded.

Male: 8   15   7   7   6   8   9   6   6   5

Female: 2   6   5   6   5   3   4   4   6   2

	Male	Female
Mean	77 ÷ 10 = 7.7	43 ÷ 10 = 4.3
Mode	6	6
Median	(7 + 7) ÷ 2 = 7	(4 + 5) ÷ 2 = 4.5
Range	15 – 5 = 10	6 – 2 = 4

> Mean suggests that males spend much longer exercising than females (7.7 vs 4.3 hours).

> Mode for both groups is the same.

> Medians also suggests that males spend longer exercising (7 vs 4.5 hours).

Looking at the **spread** of the two distributions, the presence of the outlier gives a much higher range for the males (10 vs 4 hours). Care must be taken as the majority of the males spent between 5 and 9 hours exercising and some of the females were in this range.

Comparing distributions from a table

The pay and equal rights service monitors the differences in men and women's pay. It conducted a survey of male and female salaries for physiotherapists.

For men:

Salary (£)	Frequency
$0 < £ \leqslant 15\,000$	1
$15\,000 < £ \leqslant 20\,000$	8
$20\,000 < £ \leqslant 25\,000$	20
$25\,000 < £ \leqslant 30\,000$	16
$30\,000 < £ \leqslant 50\,000$	5

Estimated mean: $1\,237\,500 \div 50 = £24\,750$

Modal group: $20\,000 < £ \leqslant 25\,000$

Median interval: $20\,000 < £ \leqslant 25\,000$ (25th/26th value will be in this group)

For women:

Salary (£)	Frequency
$0 < £ \leqslant 15\,000$	3
$15\,000 < £ \leqslant 20\,000$	5
$20\,000 < £ \leqslant 25\,000$	18
$25\,000 < £ \leqslant 30\,000$	22
$30\,000 < £ \leqslant 50\,000$	2

Estimated mean: $1\,200\,000 \div 50 = £24\,000$

Modal group: $25\,000 < £ \leqslant 30\,000$

Median interval: $20\,000 < £ \leqslant 25\,000$ (25th/26th value will be in this group)

So do women get paid less than men? Give reasons for your answer.

Comparing the two sets of data:
➤ The mean is slightly higher for men (£24750 vs £24000), which suggests they get paid more.
➤ The modal group for men is lower ($20\,000 < £ \leqslant 25\,000$ vs $25\,000 < £ \leqslant 30\,000$), which suggests women get paid more.
➤ The median for both groups is $20\,000 < £ \leqslant 25\,000$.

It is difficult to give a clear case for either but, overall, as the mean includes all the data, the survey suggests that men do get paid more than women.

Create flashcards to make sure you know the meaning of all the keywords in Modules 36 to 38. Write the keyword on one side of the flashcard and the definition on the other. Write definitions using your own words but make sure they are accurate by checking against this book.

1. Which of these can be significantly distorted by an outlier?
   Mean   Median   Mode   Range
2. The results of an analysis on the daily sales for two different stores are summarised in the table below. Which store has the best sales?

	Store A	Store B
Estimate of the mean	£3400	£3900
Modal group	$2000 < £ \leqslant 3000$	$3000 < £ \leqslant 4000$
Interval containing the median	$2000 < £ \leqslant 3000$	$2000 < £ \leqslant 3000$

Statistics Mind Map

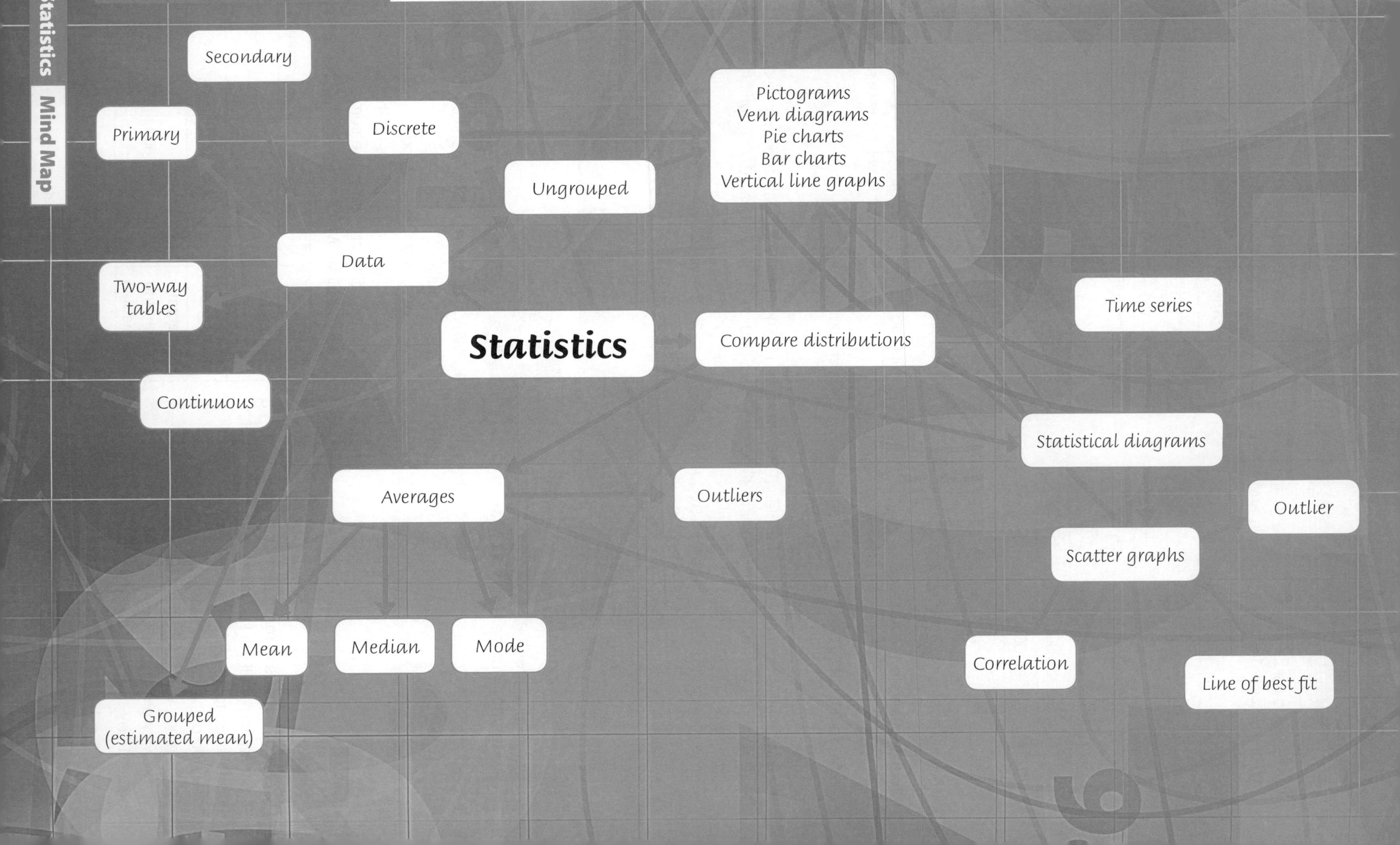
Secondary
Primary
Discrete
Pictograms
Venn diagrams
Pie charts
Bar charts
Vertical line graphs
Ungrouped
Data
Two-way tables
Statistics
Compare distributions
Time series
Continuous
Statistical diagrams
Averages
Outliers
Outlier
Scatter graphs
Mean
Median
Mode
Correlation
Line of best fit
Grouped (estimated mean)

1. In a survey of cinema-goers, 10 people said they liked Horror, 18 people said they liked Romance, 15 people said they liked Sci-Fi and 8 liked Thrillers.

   Show this data on a pictogram. [3]

2. A sales manager wants to compare the sales figures for May–October against the same period the previous year. Draw a suitable diagram to represent this data.

	Sales Year 1 (k)	Sales Year 2 (k)
May	11	6
June	17	12
July	26	18
August	34	25
September	16	12
October	5	6

   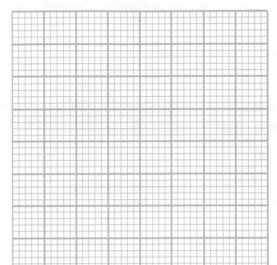

   [3]

3. Look at this data showing the weights of 200 people.

Weight (w kg)	Frequency
$10 \leqslant w < 15$	36
$15 \leqslant w < 25$	24
$25 \leqslant w < 35$	13
$35 \leqslant w < 45$	31
$45 \leqslant w < 55$	40
$55 \leqslant w < 65$	33
$65 \leqslant w < 75$	23
	200

   Find:
   (a) an estimate of the mean [4]
   (b) the median interval [1]
   (c) the modal group [1]

4. Five positive integers have a median of 5, a mean of 6 and a mode of 3.
   What is the greatest possible range for the numbers? [3]

Module 1 – Place Value and Ordering

1. (a) $+3$ (b) -8 (c) -5
2. (a) 48   (b) -8 (c) -60
3. (a) False   (b) True   (c) True
4. (a) 2.451   (b) 245 100   (c) 430

Module 2 – Factors, Multiples and Primes

1. (a) Factors of 12: 1, 2, 3, 4, 6, 12   Factors of 15: 1, 3, 5, 15
   (b) 3
2. (a) 7, 14, 21, 28, 35, 42, 49, 56   5, 10, 15, 20, 25, 30, 35, 40
   (b) 35
3. (a) $28 = 2 \times 2 \times 7$ $98 = 2 \times 7 \times 7$
   (b) $28 = 2^2 \times 7$ $98 = 2 \times 7^2$
   (c) HCF = 14   LCM = 196

Module 3 – Operations

1. (a) 18   (b) 24   (c) 38   (d) 2
2. (a) $\dfrac{5}{2}$ (b) 7   (c) $\dfrac{1}{5}$ (d) 4
3. (a) $\dfrac{1}{20}$ (b) $5\dfrac{5}{8}$ (c) $\dfrac{20}{21}$ (d) $4\dfrac{2}{3}$

Module 4 – Powers and Roots

1. (a) 4^3 (b) 3^5 (c) 2^{-3}
2. (a) (i) 3^9 (ii) 4^7 (iii) 6^7
   (b) (i) 1   (ii) $\dfrac{1}{125}$ (iii) 2
3. (a) (i) 6×10^3 (ii) 2.3×10^3 (iii) 6.78×10^{-3}
   (iv) 1.5×10^{-1}
   (b) (i) 1.8×10^{10} (ii) 2×10^6 (iii) 4.71×10^6

Module 5 – Fractions, Decimals and Percentages

1. (a) 0.6   (b) 0.875   (c) 0.6666…   (d) 0.65
2. (a) $\dfrac{9}{25}$ (b) $\dfrac{31}{125}$ (c) $\dfrac{2}{25}$
3. (a) 30.87   (b) 30   (c) 14.4

Module 6 – Approximations

1. (a) 38   (b) 370   (c) 400   (d) 5000
2. (a) 2.39   (b) 4.610   (c) 44 000   (d) 0.004 03
3. 750
4. (a) $22.5\,\text{m} \leqslant x < 23.5\,\text{m}$ (b) $4.65\,\text{kg} \leqslant x < 4.75\,\text{kg}$
   (c) $5.225\,\text{km} \leqslant x < 5.23\,\text{km}$

Module 7 – Answers Using Technology

1. (a) 12.167   (b) 4096   (c) 8
2. 3.1666…
3. (a) 9.88×10^3 (b) 7.46×10^{14}

Module 8 – Algebraic Notation

1. (a) $12ab + 2a^2b$
   (b) $9x - 9$
   (c) $2ab + 2ac + 2bc$
2. (a) g^8
   (b) p^5
   (c) $14p^2q^{-2}$
3. (a) 4
   (b) $\dfrac{1}{16}$
   (c) $\dfrac{27}{8}$

Module 9 – Algebraic Expressions

1. (a) $10 - 15x$
   (b) $x^2 - 2x - 35$
2. (a) $(x + 7)(x - 8)$
   (b) $(2p + 5q)(2p - 5q)$
3. (a) $2x^2 + x - 3$
   (b) $9x^2 - y^2$

Module 10 – Algebraic Formulae

1. (a) $p = \dfrac{b - a}{3}$
   (b) $p = \dfrac{2 - 3q}{2q - 3}$
2. (a) $p = 84$
   (b) $p = -\dfrac{3}{10}$
3. (a) $3x + x - 2 + 2x - 10 = 180$
   (b) $x = 32°$
   (c) $96°, 30°, 54°$

Module 11 – Algebraic Equations

1. (a) $x = 7$ (b) $x = \dfrac{1}{3}$
2. $x = 10, x = 1$
3. $x = 2, y = -5$
4. $x = 5$

Module 12 – Algebraic Inequalities

1. (a) $x < 8$ (b) $x \leqslant 2.5$
2. (a) $x > -3$ (b) $x \leqslant 2$
3. $x < \dfrac{8}{3}$

Module 13 – Sequences

1. 0, 3.5, 13
2. (a) $U_n = 9n - 5$
   (b) 895
   (c) No, since $9n - 5 = 779$ has no integer solution.

Module 14 – Coordinates and Linear Functions

1. $y = 3x - 2$
2. $y = 2x + 16$ and $4y = 8x - 1$
3. $y = 4x - 32$
4. $9x - 4y - 6 = 0$ or $y = \dfrac{9}{4}x - \dfrac{3}{2}$

Module 15 – Quadratic Functions

1. (a) $(-5, 0), (8, 0)$
   (b) $(0, -40)$
   (c) $x = \dfrac{3}{2}$
2. (a) $x = -5$ (b) $(-5, -65)$

Module 16 – Other Functions

1. Output $f(x) = 6, 5, 4, 3$
2. $f(x) = 2x + 5$

3. $f(x) = \dfrac{2}{x}$

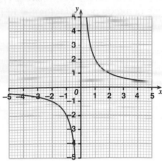

$f(x) = \dfrac{1}{x}$

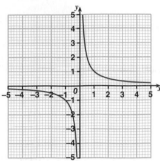

$f(x) = \dfrac{-2}{x}$

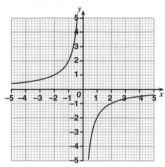

Module 17 – Problems and Graphs

1. A 2
   B 5
   C 4
   D 1
   E 3
2. **(a)** 5 seconds   **(b)** 2m/s²   **(c)** 200m

Module 18 – Converting Measurements

1. 17.8m/s (3 s.f.)   2. In UK €80 = £64 or £48 = €60
3. 84cm²   4. 2 500 000cm³

Module 19 – Scales, Diagrams and Maps

1. **(a)** 10km   **(b)** 25km   **(c)** 3.75km
   (d) 14cm   **(e)** 16cm   **(f)** 2cm
2. Accurately drawn diagrams to ±1°.
   (a) 265°   **(b)** 010°   **(c)** 120°

Module 20 – Comparing Quantities

1. **(a)** £8.40   **(b)** 392   **(c)** £19 200
2. **(a)** £19.04   **(b)** 579.12   **(c)** 8.1% (2 s.f.)
3. £384

Module 21 – Ratio

1. £16 000, $\dfrac{3}{7}$ 2. 5.4m for £4.43 (82 p/m and 86 p/m)   3. 60

Module 22 – Proportion

1. £11.55   2. 3.5   3. 15

Module 23 – Rates of Change

1. **(a)** 1   **(b)** 0.5   **(c)** $-\dfrac{3}{7}$
2. Graph 1 — C; Graph 2 — B; Graph 3 — A

Module 24 – Constructions

1. **(a)** Two lines parallel to and 3cm away from the drawn line (above and below), joined with semicircles of radius 3cm at each end
   (b) Accurate perpendicular bisector of line ST
   (c) Accurate angle bisector
2. Accurate 60° angle with visible construction lines and angle bisector

Module 25 – Angles

1. $a = 52°$ (opposite angles), $b = 52°$ (corresponding / alternate angles), $c = 112°$ (alternate to 60° + 52°)
2. 109°   **3.** $x = 76°$, $y = 130°$, $z = 130°$

Module 26 – Properties of 2D Shapes

1. Either 62°, 56° or 59°, 59°   **2.** Missing angle 240°
3. 36°   **4.** 72

Module 27 – Congruence and Similarity

1. Yes (ratios both 1.5)
2. **(a)** Yes SAS or ASA   **(b)** Yes ASA   **(c)** Yes RHS   **(d)** No

Module 28 – Transformations

1. **(a)** Translation $\begin{pmatrix} -1 \\ -5 \end{pmatrix}$
   (b) A rotation of 90° clockwise about (0, 0)
   (c) Translation $\begin{pmatrix} 0 \\ -5 \end{pmatrix}$
   (d) A reflection in line $y = 0$ (or the x-axis)
2. **(a)** A reflection in line $y = 2$
   (b) A rotation of 270° clockwise or 90° anti-clockwise about (2, 1)
   (c) A translation through $\begin{pmatrix} -7 \\ 2 \end{pmatrix}$
   (d) A rotation of 180° about (2, 10)

Module 29 – Circles

1. 25πcm²
2. 231cm
3. Semicircle (circle = 16π and semicircle = 32π)
4. $\dfrac{40}{3}$ π + 15 or 56.9cm (3 s.f.); 50π or 157cm² (3 s.f.)

Module 30 – Properties of 3D Shapes

1. **(a)** 2 faces, 1 edge, 0 vertices
   (b) 7 faces, 15 edges, 10 vertices
   (c) 6 faces, 10 edges, 6 vertices
2. Any suitable answer, e.g.

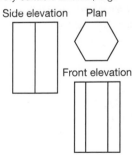

3. A sphere

Module 31 – Perimeter, Area and Volume

1. (a) 34m, 28m²   (b) 42cm, 114cm²
2. (a) 600cm³   (b) 1500π or 4712cm³ (4 s.f.)
   (c) 504π or 1583cm³ (4 s.f.)

Module 32 – Pythagoras' Theorem and Trigonometry

1. Yes. $4.2^2 + 5.6^2 = 7^2$ Pythagoras' theorem holds therefore right-angled.
2. (a) 40.6° (3 s.f.)   (b) 58.4° (3 s.f.)   (c) 36.9° (3 s.f.)
3. (a) 6.11 (3 s.f.)   (b) 3.66 (3 s.f.)   (c) 3.38 (3 s.f.)

Module 33 – Vectors

1. (a) $\begin{pmatrix} 7 \\ -2 \end{pmatrix}$ (b) $\begin{pmatrix} 9 \\ 2 \end{pmatrix}$ (c) $\begin{pmatrix} 25 \\ 1 \end{pmatrix}$ (d) $\begin{pmatrix} 7 \\ 1 \end{pmatrix}$

2. $\overrightarrow{OP} = \frac{1}{2}\mathbf{a}$, $\overrightarrow{AB} = -\mathbf{a} + \mathbf{b}$, $\overrightarrow{CR} = \frac{1}{2}(\mathbf{b} - \mathbf{c})$

Module 34 – Experimental and Theoretical Probability

1. (a) $\frac{3}{6}$ or $\frac{1}{2}$
   (b) $\frac{3}{6}$ or $\frac{1}{2}$
   (c) $\frac{4}{6}$ or $\frac{2}{3}$
2. (a) 0.54   (b) 230
3. (a) Max, because he has done more trials.
   (b) $\frac{47}{70}$

Module 35 – Representing Probability

1. (a) $\frac{1}{36}$ (b) $\frac{30}{36}$ or $\frac{5}{6}$ (c) 7
2. $\frac{4}{25}$

Module 36 – Data and Averages

1. Because it is assumed each value is at the midpoint for each group when calculating the mean.
2. No. Needs to use a bigger sample and ask a variety of boys and girls of different ages.
3. 10.53, 3h 25mins

Module 37 – Statistical Diagrams

1. (a) Any suitable answer, e.g. Slowdown in people's spending, recession, increases in prices/tax.
   (b) On an upward trend, so anything in the range 2.7–3.2 million.

Module 38 – Comparing Distributions

1. Mean and range
2. Store B as it has a higher estimated mean (£3900 vs £3400) and a higher modal group (3000 < £ ⩽ 4000 vs 2000 < £ ⩽ 3000)

Exam Practice Questions

Number

1. (a) $4 \times 58 = 232$ [1]   23.2 [1]
   (b) $23 \times 427 = 9821$ [2] (use long multiplication)   98.21 [1]
   (c) $2400 \div 8$ [1] (multiply both figures by 100) = 300 [1]
   (d) $468 \div 36$ [1] (multiply both figures by 10, then long division) = 13 [1]
   (e) −7 [1]
   (f) −3 [1]
   (g) −5 [1]
   (h) −140 [1]
   (i) 4 + 63 = 67 [1]
   (j) −31 − 9 [1] = −40 [1]

2. (a) $36 = 2 \times 2 \times 3 \times 3$ [1] (use factor trees) $= 2^2 \times 3^2$ [1]
   (b) $48 = 2 \times 2 \times 2 \times 2 \times 3$ [1] (use factor trees) $= 2^4 \times 3$ [1]
   (c) HCF $= 2 \times 2 \times 3$ [1] (the common prime factors) = 12 [1]
   (d) LCM $= 12 \times 2 \times 2 \times 3$ [1] (HCF multiplied by the remaining prime factors) = 144 [1]

3. (a) $\frac{4 \times 5}{3 \times 6}$ [1] $= \frac{20}{18} = \frac{10}{9} = 1\frac{1}{9}$ [1]
   (b) $7\frac{11}{12} - (2\frac{3}{4} + 3\frac{5}{6})$ [1] $= 7\frac{11}{12} - (6\frac{7}{12})$ [1] $= 1\frac{1}{3}$ [1]
   (c) $\frac{16}{5} \times \frac{4}{9}$ [1] $= \frac{64}{45}$ [1] (convert to improper fractions, flip second fraction and multiply) $= 1\frac{19}{45}$ [1]

4. (a) $5^{3--5} = 5^8$ [1]
   (b) $3^{(4 + 8 + 1)} = 3^{13}$ [1]
   (c) $\frac{7^7}{7^3}$ [1] $= 7^4$ [1]

5. (a) 1 [1]
   (b) $5 \times 5 \times 5 \times 5 = 625$ [1]
   (c) $\frac{1}{3^2}$ [1] $= \frac{1}{9}$ [1]

6. (a) (i) 6.7×10^7 [1]   (ii) 7×10^{-6} [1]
   (iii) 4.36×10^4 [1]   (iv) 8.03×10^{-3} [1]
   (b) (i) 12×10^{11} [1]
   (put back into standard form) $= 1.2 \times 10^{12}$ [1]
   (ii) $0.021 + 0.0057 = 0.0267$ [1]
   (put back into standard form) $= 2.67 \times 10^{-2}$ [1]
   (iii) $(1.2 \div 3) \times (10^{8-4})$ [1] $= 0.4 \times 10^4 = 4 \times 10^3$ [1]

7. (a) $5.5\,\text{m} \leqslant x < 6.5\,\text{m}$
   (b) $5.45\,\text{m} \leqslant x < 5.55\,\text{m}$

Algebra

1. (a) $3x - 6 - 4x + 4 = 0$ [1] $x = -2$ [1]
   (b) $5(x + 8) = 3(2x - 9)$ [1] $5x + 40 = 6x - 27$ [1] $x = 67$ [1]
2. A and C are parallel. [2]
3. $S = 12$ [2]
4. $6a(3a - b)$ [2]
5. (a) $x = -1.5, x = 2$ [2]   (b) (0, 6) [1]
   (c) $x = \frac{1}{4}$ [1]
6. (a) $(x + 9)(x + 7) = 0$ [1]
   $x + 9 = 0$ or $x + 7 = 0$
   $x = -9$ or $x = -7$ [2]
   (b) $(x + 10)(x - 3) = 0$ [1]
   $x + 10 = 0$ or $x - 3 = 0$
   $x = -10$ [1] or
   $x = 3$ [1]
7. 185 [2]
8. $y(x - 2) = 3x + 1$
   $yx - 2y = 3x + 1$ [1]
   $yx - 3x = 2y + 1$
   $x(y - 3) = 2y + 1$ [1]
   $x = \frac{2y + 1}{y - 3}$ [1]
9. $6x + 9y = 3$
   $6x - 4y = -36$ [2]
   $13y = 39$
   $y = 3$ [1]
   $x = -4$ [1]
10. (a)

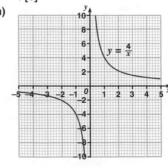

[2]

(b) Attempt to divide through by x **[1]**

$x + 1 = \dfrac{4}{x}$ **[1]**

So plot $y = x + 1$ **[1]**

Ratio, Proportion and Rates of Change

1. $2.5 \times 100 \times 100$ **[1]** $= 25\,000\,\text{cm}^2$ **[1]**
2. $2400 \div 4$ **[1]** $= 600$
   $600 \times (3 + 4 + 5)$ **[1]** $= £7200$ **[1]**
   $\dfrac{5}{12}$ **[1]**
3. $77 \div 1.22$ or $£64 \times 1.22$ **[1]**
   Game for 77 euros is cheaper **[1]** by 89p or €1.08 **[1]**
4. $\dfrac{824\,000 - 760\,000}{824\,000}$ or $\dfrac{64\,000}{824\,000}$ **[1]** $= 0.078$
   7.8% **[1]**
5. **(a)** $9 \times 200\,000$ **[1]** $= 1\,800\,000\,\text{cm} = 18\text{km}$ **[1]**
   (b) $\dfrac{18}{90} = \dfrac{1}{5}$ or 0.2 **[1]** h = 12 minutes **[1]**
6. **(a)** $\dfrac{20}{2}$ **[1]** $= 10\text{m/s}$ **[1]**
   (b) The particle is stationary **[1]**.
   (c) Line drawn from C to point (12, 0) **[1 if fully correct; 1 for any straight line from C to time axis beyond $t = 8$]**

Geometry and Measures

1. Exterior angle $180 - 156$ **[1]** $= 24°$ $\dfrac{360}{24} = 15$ **[1]**
2. Example answer: Angle $BHG = 180 - (46 + 92)$ **[1]**
   $= 42°$ (angles on straight line add up to 180°)
   Angle $HGC = 42°$ (alternate angles are equal)
   Angle $HGD = 180 - 42 = 138°$ (angles on straight line add up to 180°) **[1]**
   All reasons correct **[1]**
3. Angle $SRT = 180 - (20 + 68) = 92°$ **[1]**
   Diagram **[1]** ASA so congruent **[1]**
4. **(a)** Any rotation **[1]**, (1, −2), (1, −4), (4, −4) **[1]**
   (b) Correct reflection of the rotated image to (−1, −2), (−1, −4), (−4, −4) **[1]**
5. **(a)** **a** **[1]**   **(b)** **a** + **b** **[1]**
   (c) $\overrightarrow{CY} = \dfrac{1}{2}$ **a** **[1]**, so $\overrightarrow{OY} = \dfrac{1}{2}$ **a** + **b** **[1]**
6. **(a)** $NQ^2 + 4^2 = 6^2$ **[1]**
   $NQ^2 = 20$ **[1]**
   $NQ = 4.47\text{cm}$ **[1]**
   (b) $NQ = \sqrt{NP^2 - QP^2} = \sqrt{6^2 - 4^2}$ **[1]**
   $NQ = \sqrt{20}$ $\sin M = \dfrac{\sqrt{20}}{7}$ **[1]** $= 0.6389$
   $M = 39.7°$ **[1]**
7. Radius = 4cm **[1]**
   Volume $= \pi \times 4^2 \times 10$ **[1]**
   $= 160\pi\,\text{cm}^3$ **[1]**

Probability

1. **(a)** $\dfrac{2}{11}$ **[1]**
   (b) $\dfrac{4}{11}$ **[1]**
   (c) $\dfrac{9}{11}$ **[1]**
   (d) 0 **[1]**
2. **(a)** $1 - (0.1 + 0.2 + 0.25 + 0.1 + 0.15)$ **[1]** $= 0.2$ **[1]**
   (b) 0.15×60 **[1]** $= 9$ **[1]**
3. **(a)** HH, HT, TH, TT **[1]** $\dfrac{1}{4}$ **[1]**
   (b) HH, HT, TH, TT **[1]** $\dfrac{3}{4}$ **[1]**

4. **(a)** (D, T), (D, C), (D, H), (D, S), (N, T), (N, C), (N, H), (N, S), (P, T), (P, C), (P, H), (P, S) **[2]** **[1 if one missing or any repeated]**
   (b) 1 or 12 identified **[1]** $\dfrac{1}{12}$ **[1]**

5. **(a)**

   0.7 **[1]**   0.6, 0.4, 0.6, 0.4 **[1]**
   (b) 0.3×0.6 **[1]** $= 0.18$ **[1]**
   (c) $1 - (0.7 \times 0.4) = 1 - 0.28$ **[1]** $= 0.72$ **[1]**

6.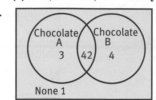

   (a) Fully correct with labels **[2]** **[1 for two numbers in the correct place]**
   (b) 1 identified **[1]** $\dfrac{1}{50}$ **[1]**
   (c) $\dfrac{3}{50}$ **[1]**

Statistics

1. Any suitable answer, e.g.

Horror	● ● ● ● ●
Romance	● ● ● ● ● ● ● ● ●
Sci-Fi	● ● ● ● ● ● ● ◖
Thrillers	● ● ● ◖

   Key: ● represents 2 people
   Suitable key **[1]**. Correctly completed pictogram for all four categories **[2]**.
2. Selecting a dual bar chart, vertical line graph or time series **[1]**
   Fully labelled **[1]**. Accurate bars/lines on chart **[1]**.
3. **(a)** Correct midpoints (12.5, 20, 30, 40, 50, 60, 70) **[1]**
      Total of 8150 **[1]**
      $8150 \div 200$ **[1]**
      Estimated mean = 40.8kg (3 s.f.) **[1]**
   (b) Median interval = $35 \leqslant w < 45$ **[1]**
   (c) Modal group = $45 \leqslant w < 55$ **[1]**
4. 3, 3, 5, 6, 13. So greatest range = 10 **[3]**